Earth and Environmental Sciences Library

The *Earth and Environmental Sciences Library* (EESL) is a multidisciplinary book series that showcases innovative approaches and in-depth reviews, aimed at enhancing the role of the Earth and Environmental Sciences communities. The series also offers valuable guidance for stakeholders, decision-makers, policymakers, international organizations, and NGOs.

Topics of interest include oceanography, the marine environment, atmospheric sciences, hydrology, soil sciences, geophysics, geology, agriculture, environmental pollution, remote sensing, climate change, water resources, and natural resource management.

To further advance these critical areas, the *Earth and Environmental Sciences Library* invites contributions from the Earth and Environmental Sciences communities in the form of edited volumes, monographs, textbooks, conference proceedings, and more. The series fosters knowledge exchange and collaboration across disciplines, encouraging comprehensive discussions and solutions to pressing environmental challenges.

André Cavalcante da Silva Batalhão ·
Pardeep Singh
Editors

Blending Approaches Towards Sustainability Focus

Springer

Editors
André Cavalcante da Silva Batalhão
Graduate Programme in Environmental Sciences
Federal University of Goiás
Goiânia, Goiás, Brazil

Pardeep Singh
Department of Environmental Studies
PGDAV College
University of Delhi
New Delhi, Delhi, India

ISSN 2730-6674 ISSN 2730-6682 (electronic)
Earth and Environmental Sciences Library
ISBN 978-3-031-93517-6 ISBN 978-3-031-93518-3 (eBook)
https://doi.org/10.1007/978-3-031-93518-3

This Springer imprint is published by the registered company Springer Nature Switzerland AG
The registered company address is: Gewerbestrasse 11, 6330 Cham, Switzerland

Foreword

Sustainability remains one of the defining challenges of the twenty-first century, necessitating interdisciplinary approaches that integrate diverse perspectives, methodologies, and frameworks. The complexity of environmental, social, and economic systems demands solutions that transcend traditional disciplinary boundaries, fostering collaboration between fields as varied as environmental science, business management, policy studies, and engineering. *Blending Approaches Towards Sustainability Focus* emerges as a timely and essential contribution to this discourse, providing a comprehensive examination of sustainability from a multidimensional perspective.

Edited by Prof. Dr. André Cavalcante da Silva Batalhão and Prof. Dr. Pardeep Singh, this volume presents an array of perceptions from leading scholars and practitioners, offering theoretical advancements, empirical research, and practical applications that highlight the necessity of an integrated approach. The editors have skillfully curated a collection of works that not only advance academic understanding but also provide policymakers and industry professionals with actionable knowledge.

Sustainability is no longer a concept that can be addressed in isolation. The pressing issues of climate change, biodiversity loss, resource depletion, and social inequalities require responses that embrace systemic thinking and collaborative action. Traditional silos of knowledge often limit our capacity to develop holistic solutions, leading to fragmented policies and inefficient interventions. This book challenges such compartmentalisation by advocating for blended approaches that synthesise multiple disciplines, ensuring that sustainability is pursued in a balanced and pragmatic manner.

The diversity of contributions in this book underscores the multifaceted nature of sustainability challenges and solutions. The chapters explore a wide range of themes, including sustainable supply chain management, environmental governance, circular economy practices, and community-based sustainability initiatives. These themes reflect the necessity of bridging corporate strategies with grassroots movements, ensuring that sustainability efforts are both globally relevant and locally contextualised.

One of the distinguishing features of this volume is its ability to bridge theoretical frameworks with practical implementation. While sustainability research often excels in conceptual depth, the translation of academic insights into policy and business practice remains a persistent challenge. This book addresses this gap by presenting case studies and empirical findings that demonstrate the feasibility of sustainability initiatives across different sectors.

Furthermore, the editors have ensured that the book maintains a balance between academic rigour and accessibility, making it a valuable resource for researchers, practitioners, policymakers, and students alike. The emphasis on evidence-based decision-making, combined with critical reflections on existing sustainability paradigms, ensures that this volume is not only informative but also transformative.

As the world faces unprecedented environmental and socio-economic disruptions, the need for cohesive and adaptive sustainability strategies has never been more pressing. *Blending Approaches Towards Sustainability Focus* offers a roadmap for navigating these complexities, encouraging interdisciplinary dialogue and fostering innovative solutions.

I am confident that this book will serve as a cornerstone for sustainability scholarship and practice, inspiring further research and action. It is my hope that readers will not only engage with its insights but also apply its principles to drive meaningful change in their respective domains.

It is with great enthusiasm that I commend this work to scholars, policymakers, business leaders, and all those committed to building a more sustainable future.

February 2025

Prof. Dr. Glaucia Aparecida Prates
Engineering and Science
Institute (ICE)
São Paulo State University (Unesp)
São Paulo, Brazil

Contents

Sustainable Development and Energy Transition

Achyutananda Mishra

Abstract Energy has been a significant driver of industrialization, economic growth, and national and global politics. At the same time, it is responsible for environmental degradation and climate change. All CO_2 emissions result from technological, economic, and demographic factors. The energy sector contributes to 90% of CO_2 emissions globally and primarily contributes to climate change. The domain of and interrelationship between energy and climate have always been contentious. Climate change engendered by global CO_2 emissions is a threat to the planet's existence. To arrest such a trend Paris Agreement (2015) calls for a global partnership limiting the atmospheric temperature to less than 2 °C above levels of pre-industrial era, and to aims at efforts limiting the increase in temperature even further to 1.5 °C. To make the energy sector sustainable SDG (Sustainable Development Goal) no 7 aims at ensuring access to affordable energy for all, i.e. sustainable, reliable and modern, which is very challenging. Further, to achieve sustainability, the world aims to achieve a net zero emission level, which requires an immediate transition from traditional energy system (fossil fuel-based) to green energy. Given the above, this chapter aims to explore the idea of energy transition and the ways and means of achieving it.

Keywords Sustainability · Net zero · Green energy · Energy Transition

1 Introduction

Energy, resource and natural environment have cause and effect relationship with each other. Energy is crucial for the socio-economic development—local, national and international and overall welfare, as well as a challenge to achieve sustainable development. Energy has been a significant driver of industrialization, economic growth, and national and global politics. At the same time, it is responsible for

A. Mishra (✉)
Christ University, Bangalore, India
e-mail: achyuta.mishra@christuniversity.in

A. Cavalcante da Silva Batalhão and P. Singh (eds.), *Blending Approaches Towards Sustainability Focus*, Earth and Environmental Sciences Library,
https://doi.org/10.1007/978-3-031-93518-3_1

environmental degradation and climate change. All CO_2 emissions result from technological, economic, and demographic factors. The energy sector contributes to 90% of CO_2 emissions globally and primarily contributes to climate change.

Energy is vital for survival of the human race, and its long-term availability from sources, which is safe, dependable, and environmentally sound, is crucial for future development of the planet. The economic implications of the energy is really disturbing as it causes climate change, due to emission of CO_2, air pollution, acidification of the environment, and risks of nuclear waste disposal; in addition the scarcity of fuel wood in developing countries [2].

Climate change, caused by human activities leading to warming of the planet, presents an existential threat to the humanity. Rising temperature has profoundly impacted the nature and natural system impacting humans like rising extreme weather conditions, floods, sea level, droughts, and loss of biodiversity across the globe; it has impacted people across the developing countries in terms of food security, migration and poverty, also at the same time affected small islands, cities, high mountain ranges and coastal areas; and severely affected several ecosystems [7]. Peace of the world depends on the safety from the climate change, and renewables are energy of peace [15].

Climate scientist underscore the urgency of transition from traditional energy system (fossil fuel-based) to green energy which is low or no-carbon based clean energy production for global economic development. A significant dimension of such a transition is that it is inequitable and unjust, and does not benefit everyone raising question about energy justice. The distributional dimension of transition has been long emphasized and articulated in various international climate agreements through the doctrines like 'intergenerational equity' and 'common but differentiated responsibilities' which are mirrored in proposals like US's 'Green New Deal' incorporating social justice and equity in climate change mitigation policy goals, and provisioning support like employment training, economic progress and overall development of the people which include both communities and individuals that may get badly affected due to such transition [4].

Such heightened concern of energy, which is unsustainable for the world, has drawn world attention calling for sustainable energy transition.

2 Problem of Energy

Accumulation of carbon dioxide due to burning of fossil fuel, land use changes and deforestation is a serious climatic problem. Increased concentration of carbon dioxide in the atmosphere contributed to the significant warming of the atmosphere [13]. Brundland [2: 174] prophetically observed that the exponential increase in the use of coal and gas will cause significant global warming due to CO_2 emission by 2020s and lack of technology to remove CO_2 emission from combustion of fossil fuel. According to IEA [11], 'the energy sector is the source of around three-quarters of greenhouse gas emissions today and holds the key to averting the worst effects of climate change, perhaps the greatest challenge humankind has faced.' Although global energy production and distribution system catalyses social and economic development, very often it conflicts with the requirements of social justice as it lacks equitable distribution and democratic participation [14]. The path adopted in the production and usage of energy across the globe is not sustainable. Millions of people don't have access to electricity resulting in dependence on traditional bioenergy (wood and charcoal) which has severe environmental, economic, and health impacts. Energy sector across the globe is primarily responsible for greenhouse gas emission which continues to increase at faster rate in comparison to any other sector [22].

Due to anthropogenic causes, the year 2017 witnessed the rise of global warming approximately upto 1 °C (likely between 0.8 and 1.2 °C) above pre-industrial levels; during the decade 2006–2015 20–40% of the global population lived in regions experiencing more than 1.5 °C temperature above pre-industrial [7]. World Meteorological Organisation [36] has alarmed the world by stating that "we will be exceeding the 1.5 °C level on a temporary basis with increasing frequency. We have already temporarily surpassed this level for individual months—and indeed as averaged over the most recent 12-month period…".

3 Sustainable Development

Sustainable development is a problem-solving technique encompassing environment, economy and equity which are not separate and exclusive but are deeply connected to each other; the fundamental idea behind the concept is to promote a process of harmonization between them while pursuing the path of development [19]. Sustainable development, which prescribes a set of goals, has a normative value for the world guiding the future course of the socio-economic development of the planet. It provides a holistic framework outlining social, economic and environmental goals, and call for a 'socially inclusive and environmentally sustainable economic growth'. Further, to achieve social, economic and environmental objectives, a fourth one, i.e. good governance must be achieved and that is integral to sustainable development [20].

According to Brundtland [2: 57], 'sustainable development is a process of change in which the exploitation of resources, the direction of investments, the orientation of technological development, and institutional change are all in harmony and enhance both current and future potential to meet human needs and aspirations.' Sustainable development requires conservation of natural/living resources, and harmonization of social, economic and ecological factors. It aims at benefiting the present generation without compromising the interest of the future generation [13]. It is a new growth paradigm which is dynamic, and seeks to harmonise social, economic and environmental concerns, and promotes sustainability, it meets the need of the present generation without compromising the ability of the future generations to meet their own requirements [2, 26]. To build inclusive and peaceful societies, and promote progress of human beings and realization human rights of all the United Nations expressed its commitment towards sustainable development by setting a set of people-centered and transformative goals and targets which are named as sustainable development goals. The United Nations aims such goals to achieve by 2030 [26]. The sustainable development framework exemplifies the society's commitment to (a) economic development, (b) social inclusion, (c) environmental sustainability and (d) good governance (including security); these are necessary for the social and individual wellbeing [21].

4 Sustainable Energy

Sustainable energy is key to achieve sustainable development. To counter the challenges of global warming and acidification of the environment, the new growth paradigm must be less energy-intensive, i.e. non-renewable fossil fuels. A safe, economically viable, and environmentally sound energy system is an imperative for a bright future and sustainable human progress [2]. Sustainable energy refers to providing of access to energy services which preserve the quality of the natural environment and its regenerative capacity. The goal of achieving sustainable energy is key to achieve few other development goals like food security, health, education, gender equality, and economic development [33].

The United Nations envisions a world of universal access to reliable, affordable, and sustainable and modern energy for all, and sustainable development goal (SDG) no. 7 aims for its achievement by 2030. In addition, it aims at substantially increasing the share of renewable energy in the global energy mix, two-fold improvement in the global energy efficiency, and soliciting international cooperation in this regard [26].

According to sustainable development goals report 2020, by 2018 789 million people world over lacked electricity, reliable and affordable energy which is critical for health facilities and 1 in 4 was not connected to electricity in few poor and growing economies; progress relating to clean cooking fuels and technologies has slowed down impacting billions of women and children's' health hindering efforts towards the realisation of Goal 7 [25].

5 Energy Transition and Energy Justice

Global climate change, increased scarcity and price of petroleum have compelled the global community to mull for a transition from the conventional carbon-based energy system to sustainable energy system. This is testified by the Kyoto Protocol of 1997 which emphasized on reduction of CO_2 (carbon dioxide) and other greenhouse gases, promotion of energy sources based on low or non-greenhouse gas. This requires a combination of strategy like, renewable energy use (solar, wind, hydroelectric, biomass, etc.), increased energy efficiency, sequestration to accompany coal use and carbon capture, and may be nuclear power [23].

Energy transition—switching from fossil fuels based energy system to energy system based on renewables—is a global trend. This transformation signifies a shifting of paradigm towards a sustainable and green energy system which has been accepted by majority of the countries across the globe. Such a transformation is evident by the acceptance of the path breaking international agreement, i.e. Paris Agreement, this was drafted by the members to the UNFCCC (United Nations Framework Convention on Climate Change) aiming at combating climate change and intensify efforts for a sustainable and low-carbon future [6, 16].

Historically speaking, energy transition is an economic and efficiency move from one form to another form of energy, e. g. wood to more efficient and economic form like coal and then coal to oil and then gas and nuclear; it means transform to more efficient form of energy which strengthens economic growth and also energy access to large number of people. But the present transition is primarily driven by the necessity to counter the problem created by the climate change and the need to decarbonise the global energy system, it is about the transformation shifting entire global energy system from carbon emitting source of energy to carbon-free sources of energy and also removing the existing carbon from the atmosphere [9].

Energy transition involves the shifting from a predominant energy resources to another or a set of energy resources like shift from whale oil to kerosene during the last part of Nineteenth Century and shift from wood to coal in the Industrial Revolution era. In the contemporary time it is envisioned as a transition from carbon rich fuels (primarily from coal) to low-carbon energy resources like wind, natural gas and solar. Energy transition engenders questions about inclusivity and distribution, which needs to be considered for effective transition, producing its set of winners and losers. Energy Justice which is considered as a branch of environmental justice in current time seeks to address such issues. It centers around the idea that all should have access to affordable, safe and sustainable energy, and be capable of leading a dignified life. Further, it seeks opportunity for all to participate and decide on all energy related decision-making process [4].

Sustainable energy transition is not only about a shift to a new and highly efficient energy system, but also about managing effectively the challenges with respect to social and environmental costs, benefits and risks. Sustainable energy transition demands an interdisciplinary understanding about the technologies advances driving the transition like, renewable energy, waste to energy, energy efficient technologies, electric vehicles and energy storage. The prominent and widely accepted emerging technologies closely related to people across the globe are renewable energy, electric vehicle, and energy storage and battery [5].

Energy transition signifies a transition to a secure, affordable, sustainable, and inclusive energy system providing remedy to global energy challenges by fostering social and business values, and striking a balance between security, equity and sustainability which is referred to as energy triangle [31]. It presupposes avoidance of socio-economic disruption [33].

Just transition is a global issue, it's about poorer communities being able to afford energy through the energy transition, it's about the redistribution of wealth in the developed countries to ensure that no one is disadvantaged, it's about a balance between developed and developing countries, it's about the commitment from developed countries to fund the transition in the developing countries to adapt and can respond to the impact of transition and to ensure that billions of disadvantaged people without access to energy are not left behind as the world inch towards a zero carbon economy [9],

Energy justice encompasses the following basic tenets [4]:

a. Distributional justice—it envisages distribution of benefits and burden throughout the population which does not burdensome population unduly or deny them access to benefits [14].
b. Procedural justice—it emphasises on the procedures of energy decision-making in the dissemination of benefits and costs of all energy services are fair and equitable, and include all those who desires to participate [24].
c. Recognition justice—it recognises the requirement of an understanding of the inequalities (ongoing and historic) in the energy system and prescribes endevours seeking reconciliation of such inequalities.
d. Restorative justice—it seeks correction and restoration of distributional, procedural and recognitional injustices.

Energy justice serves as a tool to (a) conceptualise the idea of fair and equitable distribution of costs and benefits of energy services which envisions an energy-just world that promotes welfare, happiness, freedom, equity, and due process benefiting both producers and consumers; (b) analyse energy policy and technology in the light of justice concept like human rights, virtue, utility and procedural justice, and modern conceptual thinking like freedom, welfare, posterity, and responsibility; and (c) make energy decision promoting affordability, availability, good governance, due process, intergenerational equity, intragenerational equity, sustainability, and responsibility [24].

6 Equity and Energy Transition

It is an established fact that contribution of past emissions by countries contributing to climate change has not been proportionately same, and it is also a fact that all countries are not equally capable with resources—financial and otherwise—to contribute towards solving the problem which is known as common but differentiated responsibilities and respective capabilities forming the core idea of the UNFCCC [27] and Paris Agreement 2015 [12].

The multiple crises, like geopolitical, health and economic, designated as "polycrisis" force nations to reallocate resources and initiate measures to tackle energy security and affordability issues which keep changing the frontiers of global energy transition, it has led to the growth of the thinking about "energy triangle" (equity, security and sustainability). Energy equity is rooted to the idea of affordable access to modern and clean forms of energy by all supported by Equity and inclusiveness are integral to sustained economic development. Rising fuel prices due energy market volatilities has exacerbated energy poverty and stalled energy access. A strong and resilient energy transition requires equity and inclusivity [30]. Principles of social justice demands energy equity (affordable and reliable energy) which helps reduce poverty and improve quality of life hence, should be recognized as fundamental human rights; ensuring energy equity is equally important like energy security and sustainability in the way to achieve energy transition [30]. Energy transition needs to take into account its socio-economic impact on individuals and communities, and consider people as active participants in the process [11].

7 Energy Trilemma

In the process of energy transition energy security, energy equity and environmental sustainability are of crucial significance and equally important competing priorities; they constitute a trilemma in the field of energy, and their careful management and balancing is fundamental to healthy energy systems across the globe [34, 35]. The past decade has witnessed policy thinking on energy premised on the plank of energy trilemma, i.e. sustainability (achieving decarbonisation of the energy system), equity (achieving accessibility and affordability for consumers), and energy security (guaranteeing adequate supply). The primary focus has been to harmonise between them [10].

8 Renewable Energy

Renewable energy sources have a lot of potential, and are perpetually sustainable requiring commitment in terms of research and development and needs to be prioritised in national energy programmes. Realising this, the commission urged for all out efforts to expand the scope and potential for renewable energy which would form the basis for the twenty-first century global energy structure [2]. Achieving 100% renewable energy is surest way to minimize pollution, avoid climate change and its catastrophic effects, ensure energy security, and ensure energy security. Transitioning to renewable energy is the only option left to the mankind [22].

9 Net Zero

Net zero is a policy to achieve net zero emission of CO_2, methane and other greenhouse gases in the atmosphere, and eliminate all man-made greenhouse gas emissions from the atmosphere. It is possible by reduction measures, like natural and artificial carbon sink, carbon-offsetting (carbon removal from atmosphere) or by not emitting; it is interchangeably used with the terms 'net zero carbon footprint' and 'carbon neutrality' [17]. It aims at reduction of greenhouse gas emission to near zero while the residual emissions to be re-absorbed by forests and oceans [28].

To fight climate change on an urgent basis, Article 2 (1)(a) of the Paris Agreement set the target of net zero, i.e. limiting the atmospheric temperature increase to 1.5 °C above pre-industrial levels, and Article 4 (1) set the goal for the parties to the agreement [27]. Essentially Paris Agreement requires a shift to low carbon or zero carbon technologies in energy sector [18]. It is a scientific concept seeking net-zero emission with an objective to keep the rising global temperature within a limit [8].

In order to avert the catastrophic impact of climate change and protect life on the planet, emissions level need to be reduced by 45% in 2030 and reached to net zero by 2050 [28].

Reduction of CO_2 emission to net zero globally by 2050 will help achieve the target of limiting the global average temperature to 1.5 °C, and this calls for transformative efforts in the production, transport and consumption of energy resulting in an energy system which is clean and resilient [11].

10 Private Sector Vis-à-Vis the Energy Transition

Climate change is universally acknowledged to be a mankind's common concern and promotion of access to sustainable energy universally, especially in low-income countries and in particular Africa, through increased usage of renewable energy is the need of the hour ... [27]. There is a substantial investment gap for the deployment of wind energy and solar and, private actors have a significant role in the process of energy transition [12]. There is a massive requirement of capital to install and upgrade the capacity for generating renewable energy, technologise electricity grids, revamp public and goods transport, improvise the energy efficiency in buildings [22].

11 Digitalisation and Energy Transition

Global deep emission cut can be achieved by the adoption of all technologies needed; governments must deploy available technologies and promote technologies those are not in the market and plan and incentivize all infrastructure investment, including smart transmission and distribution grids [11].

The advent of digital technology has made it possible to transform physical action into digital and significantly reduced the dependency on geographical locality, a key development in this respect is decentralized renewable energy production and the smart grid technology is an outcome of it. The smart grid technology i.e. smart meters and ICT (information and communication technologies) enabling active interface between the demand and supply sides in energy value chains. This is a decentralized electronic communication which does not require brokering by energy companies and their intermediaries. This takes place in closed circuit, i.e. a micro grid, or by using the existing distribution networks, in both cases, however, relying on smart grid technology. 'In enabling peer-to-peer trading of self-produced energy, digitalization thus has the potential to solve a core problem of distributed generation and in particular prosumership (Prosumer—a consumer who becomes involved with designing or customizing products for their own needs), that is, how to trade the energy produced and on which market' [6, 16].

According to Arrow [1], 'a smart grid is an electrical power distribution infrastructure that provides two-way communication between the utility provider and customers. Digital technologies that contribute to smart grid technology include power/current sensors, controls, data centers, and smart meters.' According to Croonenbroeck and Lowitzsch [6, 16] a micro grid is, 'a group of interconnected loads and distributed energy resources within clearly defined electrical boundaries that act as a single controllable entity with respect to the grid.

'Decentralized renewable energy (DRE) is defined as on-site, off-grid, mini-grid or distributed energy systems that use renewable energy resources including small hydro, agriculture & forest biomass waste, wind, solar, and other new renewable energy resources'. It is available locally, does not or minimally impact environment (local and global), protects natural capital, helps mitigate climate change. It is less affected by the volatility of the fossil fuel energy markets, provides local employment, technological development, and stimulates economic growth [18].

Investing in DRE is has a lot of potential benefits like

- it will accelerate the fossil fuel phase out,
- it will help achieve other SDGs such as education, health, food security, jobs, and gender,
- it will accentuate socio-economic development,
- it will involve multiple actors involving communities [29].

12 Challenges to Energy Transition

There are challenges in the path of the adoption of renewable energy and electric vehicle. For instance, solar and wind energy are intermittent and dependent on local meteorological conditions and power transmission, hence, cannot be relied upon as a steady power supply source; the operation and maintenance of the equipment used for the purpose is very challenging as their design life is shorter than their service life; limited capacities of batteries substantially reduce the operation of electric vehicles [5].

Overcoming the technological barriers requires urgent attention to promote the development of the new technologies. Energy transition goal is global and unanimous in nature but the motivation to achieve is manifold and diverse, they differ region to region, country to country and even among regions within a country, based upon the historical development of energy markets, and cultures. This heterogeneity in the motivations conflicts with global goal of energy transition and deployment of renewable energy [6, 16].

Energy transition progress gap between advanced and developing countries is widening as countries focus on energy security at the cost of equity which potentially slows down the process of energy transition. Further, disproportionate increase in energy prices affects poorer households potentially cut off their access to energy. Moreover, fossil fuel subsidies to protect consumer interest create financial burden on the government leading to misplaced priorities affecting smooth energy transition [32].

Developing countries face challenges like [3]:

1. Predominance of traditional modes of use of energy like animal waste, fuelwood, and charcoal for domestic purpose, especially in rural areas.
2. Energy poverty, i.e. lack of electricity, clean fuels; inefficiencies; and non-affordability.
3. Unreliable power systems and network.
4. Privatisation of power sector has not yielded significant gains in term of economic performance; rather it has promoted competition leading to capital utilization and labour productivity, and higher output electricity generation. Moreover, the political unwillingness to undertake market reforms or incompatibility between public and private sector leads to financial issues in power sector.
5. Rapid increase in the transport demand resulting in heavy reliance of this sector on fossil fuels which drives rapid economic growth.
6. Low level of knowledge in the citizens and their participation in the energy-related issues and inequity
7. Neo-colonisation of energy and gas resources.

12.1 War in Ukraine

Energy market has been disrupted due to invasion of Ukraine by Russia, and resultant economic sanctions on it engendering questions about the current and future role of gas in energy and power systems. It has triggered geopolitical crises prompting strategic and competing energy decision making by governments and non-state/private entities presenting a significant barrier to achieving clean energy transition. On one hand, encouraged search for the alternative gas supplies, and on the other, it has created a renewed urgency inspiring accelerated and faster energy transition [12].

13 A World Economic Forum Roadmap for Energy Transition

To realise the goal, energy transition needs to be driven by, in addition to sustainability and environmental concern, inclusivity providing energy security and access, and promoting economic growth and development. To achieve the a smooth energy transition, there is a dire need for a multi-dimensional inclusive approach reflecting a wide spectrum of perspectives and ideas, diverse policy approach recognizing various needs of a cross-section of people. The inclusive approach to achieve energy transition is five dimensional

(a) Geo-inclusivity—It recognizes that to achieve the common goal different approaches are required by different regions; the starting points within the energy triangle (equity, security and sustainability) shall be different like, for the Global South energy security is intrinsically linked to food security which is not so for the Global North (Developed Nations). This requires inclusion of the requirement of the needs of the Global South like support in term of finance, technology, raising the standard of living of all to facilitate smooth energy transition.
(b) Society and people inclusivity—This requires inclusivity within each regions and countries and calls for creation of job and reskilling those affected by energy transition. A smooth and achievable energy transition needs to recognize the needs and contributions of local people and society.
(c) Industrial Inclusivity—Tackling challenges of climate change is one of the most expensive venture in human history requiring resources, capital and technology, and in the path to zero emission industry need to develop resources, technology and create value chains to lower emission with energy technologies like solar panels, wind turbines, storage devices and all modes of transportation and infrastructure.
(d) Technology Inclusivity—Technology is key to tackle climate change problem and all technologies approach fostering research and development, innovation and collaboration is required to achieve breakthroughs in energy transition. Technology inclusivity recognizes that the resource intensive new low-emission energy technologies "a complex combination of new value and supply chains including critical minerals, metals and materials (the 3Ms)" must be sourced from different geographical locations and regions.
(c) Regulatory Inclusivity—It requires governments and society to be cautious enough to not to have exclusive policies resulting in mismatch between economies, growing populations, and energy demand, and under invest in affordable and secure energy supply. So far, Global North has been leading the narrative on net-zero with little involvement from the Global South [31].

14 Way Forward

A successful energy transition depends upon strong and vibrant enabling environment which provides an efficient regulatory framework and a climate to attract and employ capital on a large scale aligned to a comprehensive framework outlining policy for renewable energy, energy efficiency and energy access drawing a clear and credible roadmap to achieve net zero which is aided by an efficient carbon pricing [30].

To mitigate the effects of greenhouse gas emission require reduction of fossil fuel production and quick and sustained establishment of alternate and clean mode of energy production and consumption. To reach net-zero emission targets there is need of drastic reduction of oil and gas production, it must be 15% for oil and 30% for gas by 2030, and both 65% by 2025. This require immediate and large scale shift of investment towards clean energy sources. International Institute for Sustainable Development recommends the following to achieve limiting the global warming to 1.5 °C:

- Governments need to stop the development and licensing of any new gas and oil field.
- Governments need to introduce new policies to promote wind and solar capacities across the globe.
- An enabling environment should be created by the governments to redirect flow of public and private capital for clean energy transition.
- The ongoing energy crisis, especially faced by Europe should not be dealt with by increasing reliance on gas.
- Investment treaties need to be streamlined to empower the governments to enforce phasing out of the oil and gas reliance in compliance with the 1.5 °C.
- Governments require to regulate the long-term corporate goals—global and national level—to ensures compliance with the goals of Paris Agreement.
- Governments need to set up mandatory requirement norms for the reporting by companies [12].

To meet energy needs through renewable sources, the world needs to consider the following:

- Conservation of energy by enhancing energy efficiency and minimise energy wastage
- Maximisation of electricity usage generated by renewables.
- Equitable usage of energy.
- Careful planning needs to be done to generate renewable energy not only from biomass but also from land and sea space.
- Change in the lifestyle of people in the use of energy (travel and consumption).
- Rethinking in the financing system.
- Doubling of global expenditure on renewables and upgradation of energy efficiency.
- Strengthening of governance (Local, National and regional) in securing equitable energy.
- Unparalleled international cooperation and collaboration is needed in bridging the gap between the energy-poor and energy-rich nations across the globe [22].

15 Conclusion

Achieving sustainable development is the central challenge of present milieu. Climate change fuelled by greenhouse gas emission is a threat to sustainable development on many counts. World Meteorological Organisation has warned and cautioned about the tardy progress as to limiting the temperature to 1.5 °C which limits the potential of achieving net zero target. Energy transition has multiple challenges like, justice and equity challenges, development challenges, and technological challenges which affect sustainable development across the globe.

The world is passing through an unprecedented time in facing the effects (real and prospective) of climate change, at the same time it has gained unprecedented skill and ability to monitor the climate to take informed actions [36]. Shifting from fossil fuels to renewable energy sources needs technical, financial and social innovations, hence, there is need to rebuild a new energy infrastructure and individuals need to be sensitised to adopt consumption habits to synchronising their demand with the supply of energy sources [6, 16]. Energy transition on the scale and speed required by net zero pathway is not possible without continuous participation and sustained support by citizens [11].

'It is thus crucial to accelerate the deployment of DRE through supportive regulatory, financial and policy frameworks, by involving all stakeholders, and by enabling a decentralised and citizen-oriented organisation of the energy sector' [29]

Phasing out subsidies on fossil fuels will help the transition towards green energy and economy. The world must gear up to limit global warming to 1.5 °C to stop the disastrous consequences of the climate change on societies, economies and the environment. Concerted and whole-hearted efforts from all, i.e., private individuals, societies, private entities, investors, politicians et al., is the need of the hour to fulfill the goal of total renewable energy at the earliest possible time.

References

1. Arrow (2022) What is a smart grid and how does it work? https://www.arrow.com/en/research-and-events/articles/what-is-a-smart-grid-and-how-does-it-work
2. Brundtland GH (1987) Our common future. United Nations. https://digitallibrary.un.org/record/139811?v=pdf
3. Cantarero MMV (2020) Of renewable energy, energy democracy, and sustainable development: a roadmap to accelerate the energy transition in developing countries. Energy Res Soc Sci 70:101716. https://doi.org/10.1016/j.erss.2020.101716
4. Carley S, Konisky DM (2020) The justice and equity implications of the clean energy transition. Nat Energy. https://doi.org/10.1038/s41560-020-0641-6
5. Chen B, Xiong R, Li H, Sun Q, Yang J (2019) Pathways for sustainable energy transition. J Clean Prod 228:1564–1571. https://doi.org/10.1016/j.jclepro.2019.04.372
6. Croonenbroeck C, Lowitzsch J (2019) From fossil to renewable energy sources. In: Lowitzsch J (ed) Energy transition financing consumer co-ownership in renewables. Palgrave Macmillan, chap 2, pp 29–58

7. Delmotte VM, Zhai P, Pörtner H, Roberts R, Skea J, Shukla PR, Pirani A, Moufouma-Okia W, Péan C, Pidcock R, Connors S, Matthews JBR, Chen Y, Zhou X, Gomis MI, Lonnoy E, Maycock T, Tignor M, Waterfield T (2019) Global warming of 1.5 °C. Intergovernmental Panel on Climate Change. https://www.ipcc.ch/site/assets/uploads/sites/2/2022/06/SR15_Full_Report_HR.pdf
8. Fankhauser S, Smith SM, Allen M, Axelsson K, Hale T, Hepburn C, Michael Kendall J, Khosla R, Lezaun J, Mitchell-Larson E, Obersteiner M, Rajamani L, Rickaby R, Seddon N, Wetzer T (2022) Nature climate change, Vol 12. https://www.nature.com/articles/s41558-021-01245-w
9. Henderson J (2021) The energy transition explained. Oxford Energy Institute. https://www.youtube.com/watch?v=9T5c4Tgnfm4&t=75s
10. Henderson J, Meidan M (2023) Introduction—rebalancing the energy trilemma. Oxford Institute for Energy Studies. https://www.jstor.org/stable/resrep47151
11. International Energy Agency (IEA) (2021) Net Zero by 2050 a roadmap for the global energy sector. International Energy Agency
12. International Institute for Sustainable Development (IISD) (2022) Navigating energy transitions: mapping the road to 1.5 °C. https://www.iisd.org/system/files/2022-10/navigating-energy-transitions-mapping-road-to-1.5.pdf
13. International Union for Conservation of Nature and Natural Resources (IUCN), United Nations Environment Programme (UNEP), World Wildlife Fund (WWF) (1980) World conservation strategy: living resource conservation for sustainable development. www.cisdl.org://biodiversity-biosafety/public/cods/wcs.pdf
14. Jones BR, Sovacool BK, Sidortsov RV (2015) Making the ethical and philosophical case for "energy justice." Environ Ethics 37(2):145–168
15. Lazard O (2022) The blind spots of the green energy transition. https://www.youtube.com/watch?v=za6dE5JrNB0
16. Lowitzsch J (2019) Introduction: the challenge of achieving the energy transition. In: Lowitzsch J (ed) Energy transition financing consumer co-ownership in renewables. Palgrave Macmillan, chap 1, pp 1–26
17. Matemilola S, Salami HA (2020) Net zero emission. In: Idowu SO et al (eds) Encyclopedia of sustainable management. https://www.researchgate.net/publication/344224009
18. Noor R (2019) Decentralized renewable energy for improving energy access in the LDCs. https://climate.mit.edu/posts/decentralized-renewable-energy-improving-energy-access-ldcs#:~:text=DRE%20is%20defined%20as%20on-site%2C%20off-grid%2C%20mini-grid%20or,wind%2C%20solar%2C%20and%20other%20new%20renewable%20energy%20resources
19. Ruhl JB (1999) Sustainable development: a five-dimensional algorithm for environmental law. Stanford Environ Law J 18:31. https://scholarship.law.vanderbilt.edu/faculty-publications/504
20. Sachs JD (2015) The age of sustainable development. Columbia University Press
21. Sustainable Development Solutions Network (SDSN) (2012) A framework for sustainable development. http://www.jstor.com/stable/resrep16082
22. Singer S, Denruyter J-P, Yener D (2017) The energy report: 100% renewable energy by 2050. In: Towards 100% renewable energy, pp 379–383. https://doi.org/10.1007/978-3-319-45659-1_40
23. Solomon BD, Krishna K (2011) The coming sustainable energy transition: history, strategies, and outlook. Energy Policy 39:7422–7743
24. Sovacool BK, Dworkin MH (2015) Energy justice: conceptual insights and practical applications. Appl Energy 142:435–444. https://doi.org/10.1016/j.apenergy.2015.01.002
25. United Nations (UN) (2020) The sustainable development goals report. https://unstats.un.org/sdgs/report/2020/The-Sustainable-Development-Goals-Report-2020.pdf
26. United Nations (UN) (2015) Transforming our world: the 2030 agenda for sustainable development. https://www.un.org/en/development/desa/population/migration/generalassembly/docs/globalcompact/A_RES_70_1_E.pdf
27. United Nations Framework Convention on Climate Change (UNFCCC) (2016) The Paris Agreement. https://unfccc.int/sites/default/files/resource/parisagreement_publication.pdf

28. United Nations (2022) For a livable climate: net-zero commitments must be backed by credible action. https://www.un.org/en/climatechange/net-zero-coalition
29. United Nations (2019) Decentralized renewable energy for access. https://sdgs.un.org/sites/default/files/2021-05/ACTION%20BRIEF%205%20-%20DECENTRALIZED%20RENEWABLE%20ENERGY%20FOR%20ACCESS.pdf
30. World Economic Forum (WEF) (2023a) Fostering effective energy transition. https://www3.weforum.org/docs/WEF_Fostering_Effective_Energy_Transition_2023.pdf
31. World Economic Forum (WEF) (2023b) How to create a framework for achieving an inclusive energy transition. https://www.weforum.org/agenda/2023/10/framework-for-achieving-an-inclusive-energy-transition/
32. World Economic Forum (WEF) (2023c) What is energy equity and why is it stalling globally? https://www.weforum.org/agenda/2023/07/energy-equity-stalling-report/
33. Weijnen M, Lukszo Z, Farahan S (2021) Introduction. In: Weijnen MPC, Lukszo Z, Samira F (eds) Shaping an inclusive energy transition. Springer, chap 1, 1–11
34. World Energy Council (n.d.) World energy trilemma framework. Retrieved 26 May 2024, from https://www.worldenergy.org/transition-toolkit/world-energy-trilemma-framework
35. World Energy Council (2019) World energy trilemma index, 2019. https://www.worldenergy.org/assets/downloads/WETrilemma_2019_Full_Report_v4_pages.pdf?v=1601288224
36. World Meteorological Organisation (2024) Global temperature is likely to exceed 1.5 °C above pre-industrial level temporarily in next 5 years. https://wmo.int/news/media-centre/global-temperature-likely-exceed-15degc-above-pre-industrial-level-temporarily-next-5-years

Financing and Investment in a Circular Economy: Paving the Way for Sustainable Growth

Shikha Daga, Kishore Kumar, Kiran Yadav, and Pardeep Singh

Abstract Environmental deterioration and drastic climate change have forced economies worldwide to make a transition towards green growth by adopting practices of circular economy. The paradigm shift towards the adoption of the circular economy concepts however necessitates the use of contemporary financial instruments to facilitate strategic reforms. The chapter provides a detailed analysis of the varied financial and investment needs, hurdles, and blockages in its adoption and the future prospects that lead to the adoption of sustainable practices thereby promoting growth that not only generates wealth but is also sustainable and green. The chapter highlights the concept of a circular economy in the globalized world along with highlighting the financial and investment incentives that promote green growth. The chapter also provides varied measures that can help in effective waste management along with building a sound infrastructure thereby enhancing resource efficiency and satisfying the multiple goals addressing the three P's—People, Planet, and Profit. The chapter emphasizes the significance of transitioning from the conventional linear 'take-make-dispose' paradigm to a regenerative and restorative approach to production and consumption.

Keywords Circular economy · Sustainable development · Green economy · Inclusive growth

1 Introduction

Rising environmental degradation, increased greenhouse gas emissions, and destruction of the natural flora and fauna have led to the rising interest in contemporary techniques, models and technology to ensure that there is development but not at the

S. Daga (✉) · K. Yadav · P. Singh
PGDAV College, University of Delhi, Delhi, India
e-mail: shikha.menani@pgdav.du.ac.in

K. Kumar
School of Management, Bennett University, Greater Noida, India

A. Cavalcante da Silva Batalhão and P. Singh (eds.), *Blending Approaches Towards Sustainability Focus*, Earth and Environmental Sciences Library,
https://doi.org/10.1007/978-3-031-93518-3_2

cost of the environment [14]. There have been factors that can be attributed to rising greenhouse gas emissions and a decline in environmental conditions.

The paradigm shift towards a circular economy from the traditional linear model ensures that there is waste minimization and efficient resource utilization. Circular Economy ensures that the resources are used for their highest value throughout their service life, and then recovered and regenerated after each cycle (Fig. 1). The rising concern towards environmental degradation has made the adoption of circular economy concepts mandatory both at the micro and macro levels [9]. In order to achieve and align with the goals of the circular economy, it is crucial to introduce creative finance and investment methods that encourage the growth of circular business practices in a sustainable manner [3]. The circular economy focuses on the 3 R's namely reducing, reusing, and recycling thereby ensuring that every element in the supply chain is utilized efficiently with minimized wastage and promotion of sustainability [8].

Circular economy is an important strategy that focuses on resource efficiency along with making a shift from conventional supply chains towards more efficient and green supply chains. However, the shift necessitates infusion of money at different stages along with promoting the reuse, recycling, and refurbishment of the materials involved in the entire supply chain. The change though important and crucial for promoting a greener tomorrow makes it necessary to adopt solutions that are financially constructed in order to satisfy a variety of specific and individual requirements.

The chapter addresses the important concept of financial engineering that is crucial in adoption of the circular practices across the world. The chapter highlights varied concepts like the concept of green finance that helps in satisfying the capital requirements for wealth generation along with promoting what is known as sustainable growth via the adoption of circular practices. In addition to discussing the contemporary sustainable financial tools, the chapter elucidates the recent developments in the financial sector that can lead to the adoption of circular practices. The chapter thus would be of immense use for practitioners, academicians, researchers and policy makers that focus on fostering a green environment without hampering the growth.

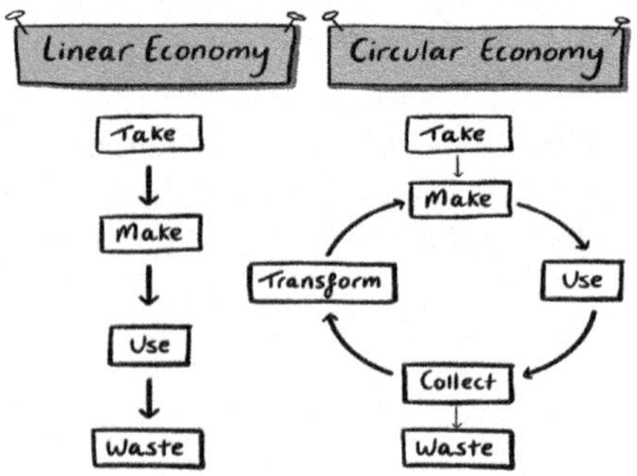

Fig. 1 Linear and circular economy: a depiction. *Source* commons.m.wikimedia.org

2 Bibliometric Review of Literature

A detailed examination of the existing literature helps in identifying the current status, gaps and scope of future research in a particular domain. A new technique that helps in reducing the bias in the extraction of literature along with providing a comprehensive analysis of the past literature is the Bibliometric analysis. It offers a systematic approach to doing a thorough examination of the current body of literature [6]. The study made use of the SCOPUS database wherein then performance analysis and scientific mapping was done to identify the most frequently used keywords, most prominent authors working in the specific domain, country wise analysis along with a detailed citation analysis that helps in identifying the most relevant research in the domain.

The following string is used to extract the relevant literature:

TITLE-ABS-KEY ("Financ*" AND "Circular Economy" AND "Sustainabl*")

It provides a dataset of 821 research documents whose analysis is provided in the present section.

2.1 Publications Per Year

Rising carbon emissions, environmental degradation, and severe climate change have forced academicians, policymakers, industrialists, and the individual at large to think of innovative techniques to contribute towards mitigating environmental deterioration [10]. Efforts made at both the macro and local levels to transition from conventional economic growth to sustainable development have also intensified study on the latter as can be seen from Fig. 2a which provides rising research interest in the area since the turn of the century. However, the importance of financing and the role of a circular economy in the promotion of sustainable development has been the area of research only recently as is visible in Fig. 2b.

2.2 Documents Per Year by Source

Research on the influence of finance and circular concepts in promoting green growth has been published in various journals across different fields. However, a larger number of documents, in terms of quantity, are found in environment-related journals such as Sustainability, Journal of Cleaner Production, Energies, and others. The issue at hand is that while the research focuses on financing aspects in promoting sustainable development, it is published in top-tier journals in the environment-related field rather than the finance field (Fig. 3). This indicates a bias in the finance

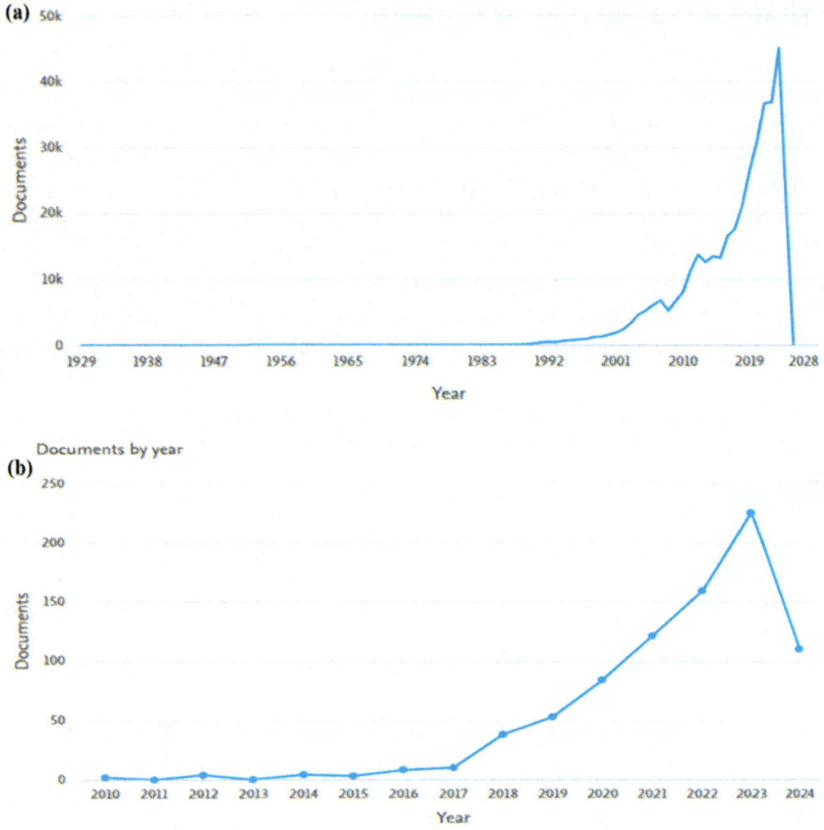

Fig. 2 **a** Documents per year on sustainable development. **b** Documents per year on financing, circular economy and sustainable development. *Source* Scopus

journals towards promoting core finance research, highlighting the need to address this gap [12].

2.3 Documents Per Year by Type

Research in the area of financing, circular economy and sustainable development is not limited to published articles but also covers edited work, short surveys, books, review papers, conference papers and book chapters showing the diversity of research in the area (Fig. 4).

A three-field plot providing the interconnectedness between the most influential authors' countries, most productive authors, and most frequently used keywords is depicted in Fig. 5. The chart depicts the most active countries in the research domain

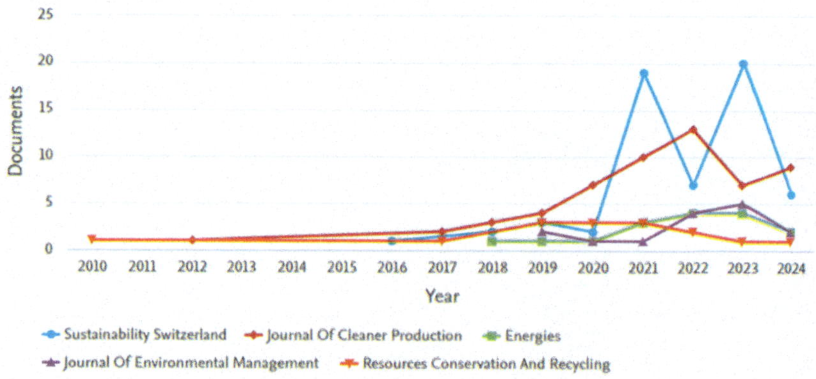

Fig. 3 Documents per year by source on financing, circular economy and sustainable development. *Source* SCOPUS

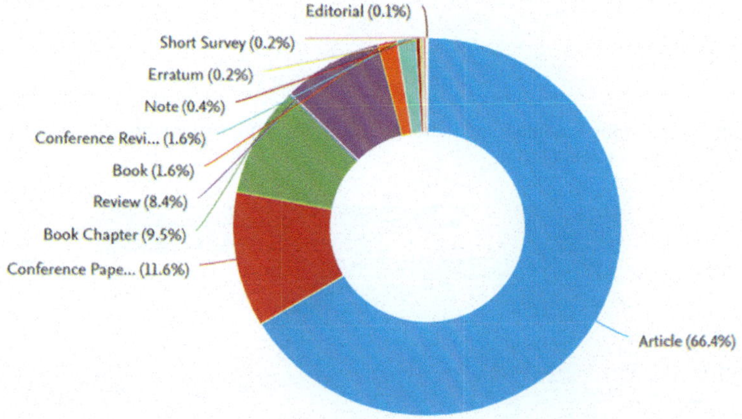

Fig. 4 Documents by type on financing, circular economy and sustainable development. *Source* SCOPUS

as emerging economies of Italy, and India followed by developed countries of the United Kingdom and the United States showing how the role of finance and promoting circular economy is of utmost importance in the economies that are growing at a rapid pace. The most influential keywords are found to be circular economy, sustainable development, green finance and climate change. An important point to address here however is whether it be solving the global issue of climate change or promotion of sustainable development, collaboration be at micro or macro level is of utmost importance. The case is the same with the research and is highlighted in Figs. 6 and 7 which provides the level of collaborations and co-citations in the field of study. The analysis is important as it provides a glimpse of the countries that are working in close connectivity to tackle a common issue.

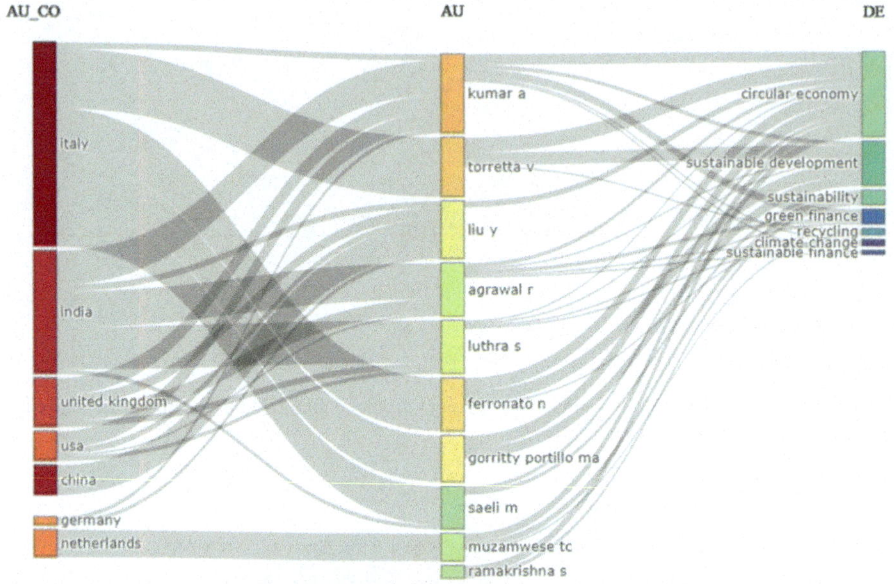

Fig. 5 Three field plot on financing, circular economy and sustainable development. *Source* Biblioshiny, Author's Computation

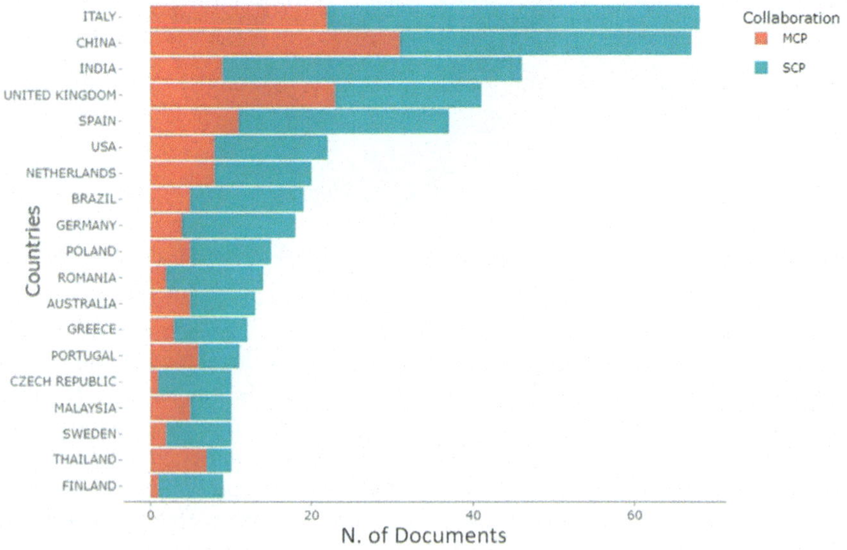

Fig. 6 Most relevant affiliations in the field of financing, circular economy and sustainable development. *Source* Biblioshiny, Author's Computation

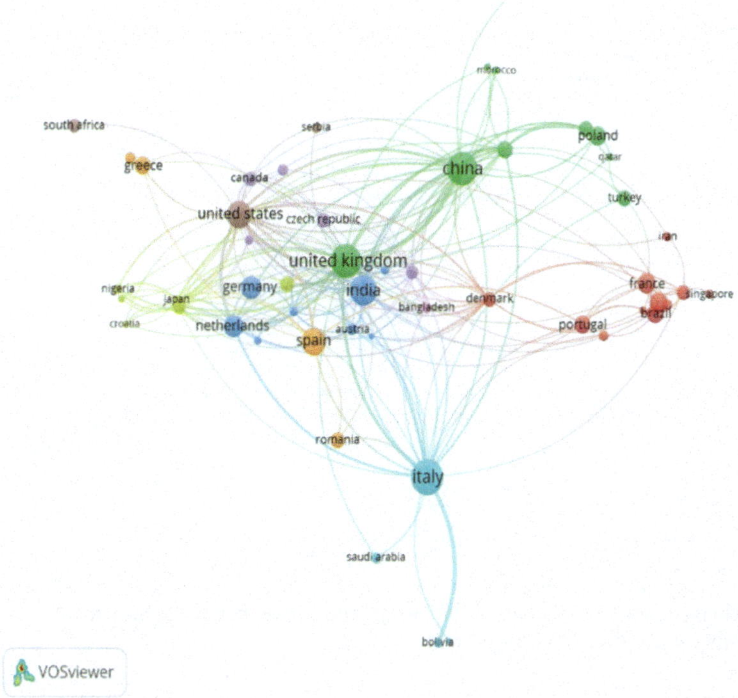

Fig. 7 Co-citations in the field of financing, circular economy and sustainable development. *Source* VOSviewer, Author's Computation

2.4 Thematic Evolution

Figure 8 subdivides all the research in the domain of Financing, Circular Economy and Sustainable Development into four quadrants based on its centrality and density. The four quadrant figure thus provides an idea about the sub-areas that are over-researched and those where still there is a gap and policy formulation is mandated. It thus gives a quick glimpse to the researcher about the area that still needs extensive research and solution.

3 Circular Economy and Resource Extraction

Circular economy is pertinent for attaining the sustainable development goals at the economy level along with promoting resource efficiency at the micro level [3]. Economies worldwide today are faced with the situation of mass resource extraction that not only adds to their carbon footprints but also make them vulnerable to resource extinction. Over the last fifty years there has been a structural shift in the

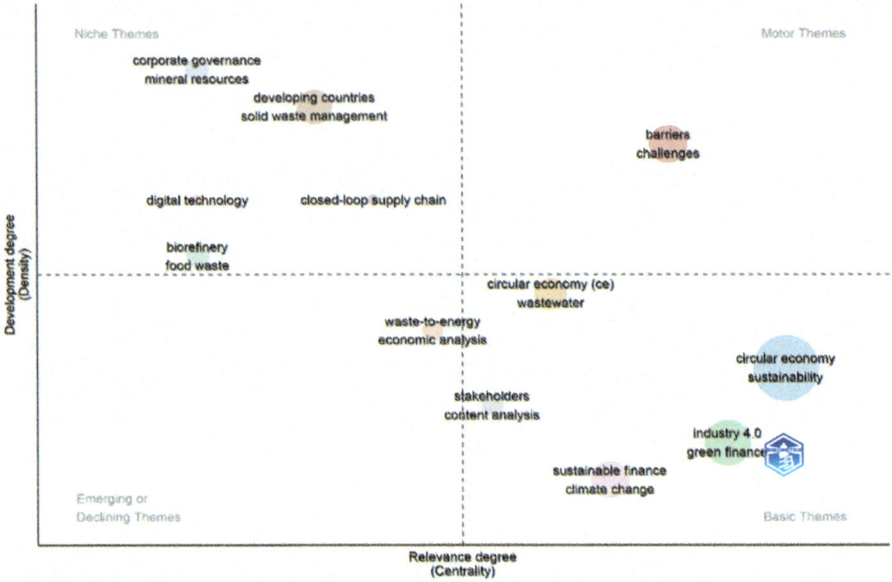

Fig. 8 Thematic map in the field of financing, circular economy and sustainable development. *Source* Biblioshiny, Author's Computation

national income lost due to energy and natural resource depletion and the carbon emission that has made circular economy gain momentum worldwide [11]. The break is visible during 1990s when the world realized the importance of attaining sustainable development along with generation (Table 1). The circular economy is intrinsically linked to resource extraction as it aims to decouple the economic growth from the former. It is not only concerned with reducing the environmental impact of the production and consumption process but also aims to achieve sustainability and attain economic resilience [13]. The linear economy focuses on resource extraction to generate wealth without any focus on waste generation and its use. Linear economy output is thus characterized by three aspects namely Resource depletion, Environmental Degradation and Waste Generation (Fig. 9).

Circular economy on the other extreme is a contemporary counterpart of the linear economy that integrates economic development, and output generation along with waste minimization and resource efficiency (Fig. 10).

The circular economy thus not only adds to the promotion of sustainable growth but also helps in reducing the demand for resource extraction and promoting a green resilient future [1]. The solution however is not straightforward as the transition towards a circular economy has a wide array of barriers and limitations which is discussed in the next section.

Table 1 Global natural resource depletion and carbon emission

Year	Adjusted savings: energy depletion (% of GNI)	Adjusted savings: natural resources depletion (% of GNI)	CO_2 emissions (kg per 2015 US$ of GDP)
1974	2.0048	2.3046	–
1975	1.7301	1.9419	–
1976	1.8726	2.0693	–
1977	1.8295	2.0609	–
1978	1.6233	1.7987	–
1979	3.0611	3.2885	–
1980	3.4068	3.7035	–
1981	2.6296	2.8620	–
1982	1.6190	1.8674	–
1983	1.7875	2.0135	–
1984	1.7511	1.9149	–
1985	1.5823	1.7336	–
1986	0.6206	0.7589	–
1987	0.7976	0.9552	–
1988	0.6419	0.7794	–
1989	0.8984	0.9867	–
1990	1.0759	1.1153	0.5911
1991	0.6168	0.7113	0.5874
1992	0.6106	0.7602	0.5743
1993	0.6471	0.7692	0.5676
1994	0.5538	0.6937	0.5530
1995	0.5465	0.6982	0.5519
1996	0.7113	0.8480	0.5443
1997	0.6319	0.7563	0.5336
1998	0.3395	0.4527	0.5224
1999	0.5290	0.6264	0.5080
2000	1.0783	1.1863	0.5015
2001	0.8986	0.9872	0.4989
2002	0.8051	0.8986	0.4946
2003	0.9596	1.0713	0.5016
2004	1.1772	1.3002	0.5022
2005	1.5429	1.7097	0.5012
2006	1.5364	1.8054	0.4957
2007	1.4560	1.8348	0.4930
2008	2.1050	2.4884	0.4864
2009	1.0739	1.3187	0.4868

(continued)

Table 1 (continued)

Year	Adjusted savings: energy depletion (% of GNI)	Adjusted savings: natural resources depletion (% of GNI)	CO_2 emissions (kg per 2015 US$ of GDP)
2010	1.3224	1.7525	0.4943
2011	1.8193	2.3514	0.4931
2012	1.6041	1.9210	0.4856
2013	1.4354	1.7142	0.4817
2014	1.2694	1.5030	0.4693
2015	0.6400	0.8142	0.4528
2016	0.5342	0.7236	0.4415
2017	0.6980	0.8971	0.4337
2018	0.9899	1.1847	0.4305
2019	0.8484	1.0425	0.4187
2020	0.4715	0.6905	0.4086
2021	0.9973	1.5186	.

Source databank.worldbank.org

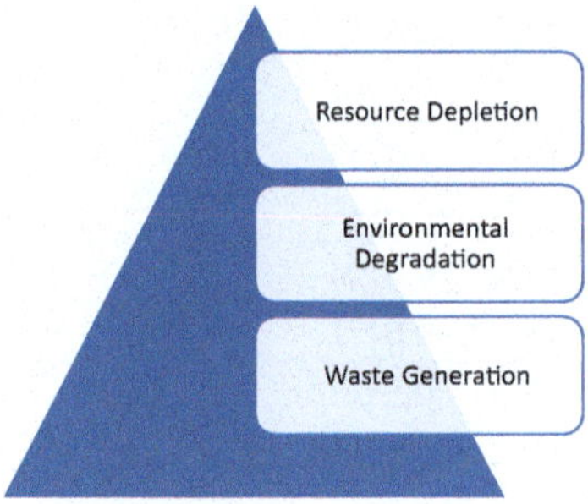

Fig. 9 Key issues of the linear economy. *Source* Author's Creation

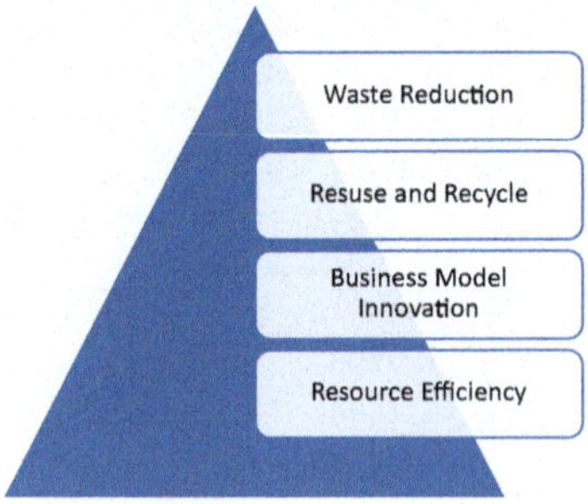

Fig. 10 Key issues of the circular economy. *Source* Author's Creation

4 Breaking Barriers: Addressing Hindrances to a Circular Economy

Identification of Circular economy as an instrument to promote sustainability is now widely accepted. However, it has not brought significant differences in the hindrances that mar the attainment of circular principles. A few such hindrances as provided by the study of Araujo Galvão et al. [2] are provided in this section.

4.1 Technological Barriers

A shift towards circular economy needs the adoption of advanced technology that can help in recycling, waste minimization and resource efficiency. However, not all countries and industries possess technological abundance that can act as a hindrance to the attainment of circular economy objectives.

4.2 Financial and Economic Barriers

Not only is the technical backwardness a prominent hindrance especially for the low-income countries but the financial requirements to make a switch is equally disturbing. Advanced technology generally comes at an accelerated cost which can mar the economies with credit crunch to take the jump.

4.3 Managerial Barriers

Adoption of circular economy principles also require a change in the mindset of the people in charge of the process. It acts as a major hindrance if the people at the top lack interest or don't have enough knowledge about the adoption of circular practices. Other than this sometimes in old organisations the working style is so imbibed that there is an inertia to change as it requires a major overhaul in the working practices.

4.4 Performance Indicator Barriers

Adoption of circular economy practices takes time to deliver and indicate results. Secondly, there is a lack of clear performance metrics that makes measuring the success of circular economy initiatives challenging due to the absence of standardized indicators, making it difficult to track progress and demonstrate tangible benefits.

4.5 Customer-Related Barriers

Demand for circular products and services can arise only when customers are fully aware and conscious about using them. Limited knowledge and awareness amongst customers at the consumption side hence act as a significant barrier to the adoption of circular technology at the production front thereby creating an obstacle to its adoption.

4.6 Cultural and Social Barriers

A circular economy can produce a hundred percent result if it is adopted at the global level. However, cross-country cultural barriers, lack of social awareness, economic inequality and the race to outperform others in terms of economic development act as a strong barrier towards promoting circular economy.

Gaining knowledge about the significance of the circular economy, its current and future advantages for the global economy, the obstacles that impede its implementation, and the remedies for these barriers will facilitate the promotion of the circular economy at both the individual and societal levels. One of the major hindrances in the circular economy is the lack of financial incentives and that is the topic of discussion in the next section.

5 Financial and Investment Incentives for Circular Economy

In addition to the various barriers that hinder the adoption of a circular economy, lack of finances and investment incentives is at the core. A few incentives that can be provided to the individuals as well as the organisations are provided in this section.

5.1 Subsidies and Grants

To ensure that modern-day firms switch to circular economy techniques, it is necessary to provide them with grants and/or incentives to ensure that the transition is voluntary and quick. Varied incentives like helping the firms in the research and development cost, adoption of cleaner technology, educating the society about sustainable consumption can help in the adoption and promotion of circular goods, services, and processes. The incentives should not only be provided at the production side but should also be introduced at the consumption side to ensure that the whole supply chain is made circular and green. This can be done by providing subsidies for circular goods consumption. It can therefore lead to sustainable and resilient development.

5.2 Tax Incentives

In addition to providing subsidies and incentives it is equally important and motivating to provide tax relief to producers and consumers who make a switch towards sustainable practices adoption. It can be in the form of reduced tax on circular products or tax credit. Other than this, at the onset tax benefits can be provided for purchasing refurbishment machines. All such incentives will thus help in making the whole supply chain green from the producers up to the consumers.

5.3 Low-Interest Loans and Green Bonds

Providing access to easy finance for promoting a circular economy is necessary to induce more and more people to adopt it. Easy access and enhanced liquidity to the green bond market can help the organizations to raise funds for sustainable practices easily at a lower cost thereby committing themselves towards green development. Similarly, the availability of low-interest-rate loans by financial institutions and the government can reduce the overall cost of transitioning towards a circular economy.

For the circular economy to develop and economic practices to become sustainable, financial and investment incentives play a major role. These incentives reduce financial impediments as well as promote innovation thereby enabling individuals and enterprises to embrace circular economy practices that save resources, lower waste creation and earn economic benefits. This is through subsidies, tax incentives, low-interest loans, green bonds and infrastructure investments that can enable governments and organizations to build a healthier and more sustainable global economy.

However, realizing the potential of these financial and investment incentives while at the same time meeting the challenges calls for efforts from all players including governments, businesses, and investors' communities among others. By utilizing these monetary rewards plus investing in a circular economy model, the world will be able to achieve sustainable development in the long run leading to a better future for all.

6 Conclusion

A move towards the circular economy is one of the strategic initiatives to promote sustainable development and bring a green and resilient future for all [4]. The objective however is not easy to obtain as there are various obstacles and hindrances in the path of success. Cross-country differences and financial deficits make it difficult for the adoption of the policy at the global level [5]. What is required to take the planet towards a sustainable path is to reduce wastage, recycle products and make them stay in the circular flow as far as possible. This is possible only when there is an awareness of its need and a self-conscious move towards its adoption. Eco-innovation is thus the key that can help the economies in taking the leap [7]. It will not only assist in reducing environmental degradation but will also promote resource efficiency and hence assist in cost reduction and resilient and green growth [9]. At the micro and organizational level, the adoption of a circular economy will help in reaching sustainable results along with reducing the ecological impact [3]. The success however of all the steps undertaken will be highly dependent on the collaborative efforts of diverse stakeholders starting from the government at the apex level and the communities at the local level. Ultimately, transitioning to a circular economy is a shared effort that can lead to lasting sustainable growth and a stronger, more prosperous future for everyone. By making smart financial investments and using innovative funding methods, we can ensure to create a circular economy that benefits both people and the environment.

References

1. Aithal S, Aithal PS (2023) Importance of circular economy for resource optimization in various industry sectors—a review-based opportunity analysis. Int J Appl Eng Manag Lett 1:191–215. https://doi.org/10.47992/ijaeml.2581.7000.0182
2. Araujo Galvão GD et al (2018) Circular economy: overview of barriers. Procedia CIRP 73:79–85. https://doi.org/10.1016/j.procir.2018.04.011
3. Barros MV et al (2021) Circular economy as a driver to sustainable businesses. Clean Environ Syst 2:100006. https://doi.org/10.1016/j.cesys.2020.100006
4. Bianchi M, Cordella M (2023) Does circular economy mitigate the extraction of natural resources? Empirical evidence based on analysis of 28 European economies over the past decade. Ecol Econ 203:107607. https://doi.org/10.1016/j.ecolecon.2022.107607
5. Domenech T, Bahn-Walkowiak B (2019) Transition towards a resource efficient circular economy in Europe: policy lessons from the EU and the member states. Ecol Econ 155:7–19. https://doi.org/10.1016/j.ecolecon.2017.11.001
6. Donthu N et al (2021) How to conduct a bibliometric analysis: an overview and guidelines. J Bus Res 133:285–296. https://doi.org/10.1016/j.jbusres.2021.04.070
7. de Jesus A, Mendonça S (2018) Lost in transition? Drivers and barriers in the eco-innovation road to the circular economy. Ecol Econ 145:75–89. https://doi.org/10.1016/j.ecolecon.2017.08.001
8. Kristensen HS, Mosgaard MA (2020) A review of micro level indicators for a circular economy—moving away from the three dimensions of sustainability? J Clean Prod 243:118531. https://doi.org/10.1016/j.jclepro.2019.118531

9. Lehmann C et al (2022) Leveraging the circular economy: investment and innovation as drivers. J Clean Prod 360. https://doi.org/10.1016/j.jclepro.2022.132146
10. Sadiq M et al (2024) Role of fintech, green finance, and natural resource rents in sustainable climate change in China. Mediating role of environmental regulations and government interventions in the pre-post COVID eras. Resour Policy 88:104494. https://doi.org/10.1016/j.resourpol.2023.104494
11. Savini F (2023) Futures of the social metabolism: degrowth, circular economy and the value of waste. Futures 150:103180. https://doi.org/10.1016/j.futures.2023.103180
12. Senadheera SS et al (2022) The development of research on environmental, social, and governance (ESG): a bibliometric analysis. Sustain Environ 8(1):1–15. https://doi.org/10.1080/27658511.2022.2125869
13. Velenturf APM et al (2019) Circular economy and the matter of integrated resources. Sci Total Environ 689:963–969. https://doi.org/10.1016/j.scitotenv.2019.06.449
14. Zhao Y et al (2024) Racing towards zero carbon: unraveling the interplay between natural resource rents, green innovation, geopolitical risk and environmental pollution in BRICS countries. Resour Policy 88:104379. https://doi.org/10.1016/j.resourpol.2023.104379
15. https://commons.m.wikimedia.org/wiki/File:Linear_versus_circular_economy.png#filelinks
16. https://databank.worldbank.org/source/world-development-indicators

Waste Printed Circuit Boards (WPCBs) of Mobile Phones in Hydrometallurgical Route: A Systematic Review

O. Saba-Bawazir, Hussien A. Kamal, Osama Bin-AlShaibah, M. Safa-Gamal, and André Cavalcante da Silva Batalhão

Abstract Due to the lockdown and restrictions imposed to confront COVID-19 pandemic over the past, mobile phone sales spiked to reach 7.1 billion in 2021 which predicts in a massive increase of waste from mobile phones and subsequently an increase of waste printed circuit boards (WPCBs) associated with mobile phones. WPCBs contain a variety of precious and base metals such as copper, gold, nickel, zinc and many more. The recycling of these WPCBs for the recovery of those metals is extremely attractive from economic and environmental perspectives. Published studies shows that hydrometallurgical process is economical and less hazardous to the environment in recycling WPCBs compared to other techniques. This review summarises and analyses the research results of hydrometallurgical process for the recycling of WPCBs from mobile phones in recent years with focusing on two phases of the hydrometallurgical route, leaching and purification. We conducted a systematic search in three electronic databases of original full-length articles, namely Web of Science, Science Direct, and Scopus, for studies related to recycling WPCBs from

O. Saba-Bawazir (✉)
School of Engineering, Faculty of Science and Technology, Centre for Waste and Resource Management, University of Central Lancashire, Preston, UK
e-mail: Sababawazir@yahoo.com

Chemical Safety & Waste Management Department, Environmental Protection Authority, Ministry of Water and Environment, Aden, Republic of Yemen

H. A. Kamal
Chemistry Department, Faculty of Science, Al-Azhar University, Assuit, Egypt

O. Bin-AlShaibah
Institute of Biological Sciences, Faculty of Science, Universiti Malaya, Kuala Lumpur, Malaysia

M. Safa-Gamal
Department of Chemistry, Faculty of Science, Universiti Putra Malaysia, UPMSerdang, Selangor, Malaysia

Technical Department, Aden Refinery Company, Aden, Yemen

A. C. da Silva Batalhão
Graduate Programme in Environmental Sciences, Federal University of Goiás, Goiânia, Goiás, Brazil

A. Cavalcante da Silva Batalhão and P. Singh (eds.), *Blending Approaches Towards Sustainability Focus*, Earth and Environmental Sciences Library,
https://doi.org/10.1007/978-3-031-93518-3_3

mobile phones by using a hydrometallurgical process between 2015 and 2021. In total, we found 129 records. For eligibility, 121 full-text publications were reviewed; 120 were excluded, resulting in the inclusion of 9 published studies that matched the predetermined eligibility criteria. This review concludes with recommendations for future research on the selection of optimal leaching and purification techniques for metal recovery from WPCBs particularly of mobile phones.

Keywords Waste printed circuit boards (WPCBs) · Mobile phones · Hydrometallurgical process · Metals recovery

1 Introduction

Our dependence on electrical and electronic equipment (EEE) is growing day-after-day due to the significant technological advancement in all disciplines, that has led to a decrease in the lifespan of electronic gadgets and a huge increase in the generation of electronic waste [30, 36, 40, 106, 118]. The United States Environmental Protection Agency (EPA) estimates that the volume of electronic waste is currently growing faster than it has ever been and has exceeded hazardous waste streams [105], and municipal solid waste such as food, wood, plastic, and household waste etc. [37, 48, 123]. According to Global E-waste Monitor 2020 report, about 53.6×10^6 Mt of e-waste were generated in 2019 with an estimated growth to 74.7×10^6 Mt by 2030 worldwide [24]. E-waste consists of a mixture of precious metals (Au, Ag, Pd, Pt, etc.), base metals (Cu, Zn, Ni, Al, etc.) in addition to some hazardous metals (Hg, Cd, Be and Pb) [49, 108], which may have serious environmental and health impacts if not managed properly [57]. The value of these metals was estimated to be 57 billion USD in 2019 [24].

E-waste from mobile phones is known as a precious secondary resource of metals as they are rich in metal content of precious, base and critical elements with ever growing volume production due to their short lifespan [3, 76]. The associated waste resulted by mobile phones high usage is seen as a critical concern for the environment [43]. Additionally, waste from mobile phones provide the largest share number of e-waste [5, 120], due to the large increase in their stockpiles and their low collection rates [82]. Currently, e-waste from mobile phones presents a larger ownership per capita in comparison with other categories of e-waste [103]. For instance, it's estimated that approximately 400 million units of mobile phones are categorized as waste globally [59]. Statistics showed that the number of mobile phone users was 7.1 billion in 2021 and expected to reach 7.49 billion in 2025 worldwide [102]. Moreover, COVID-19 restrictions that led people to stay and work from home resulted in an increase in electronic devices sales particularly portable electronic devices such as mobile phones [85, 101].

Most electronic devices contain printed circuit boards as an integral part despite of their sizes [63]. Mobile phones consist of screen display, casing shell, battery, and printed circuit boards (PCBs) with the latter being rich in precious and base metal

Table 1 WPCB contents in several types of e-waste

Type of E-waste	PCB contents (%)
Mobile phone	26
Computer	20
Television	9.5
Washing machine	3
Air conditioner	1
Refrigerator	0.5

such as Cu, Ag, Fe, Ni, Sn, Au, and Pb [44], compared to other small-sized e-waste devices. Although PCBs compromise only 9–10% of total mobile phones weight [6], the metals contained in the PCBs make up to 80% of the mobile phones value [31]. Printed Circuit Boards (PCBs) are additionally composed of significant quantities of heavy metals, polyvinyl chloride polymers, and brominated flame retardants [63]. Additionally, hazardous metals such as lead (Pb) and mercury (Hg) and some others from PCBs have a negative impact on the environment [114].

PCBs comprise concentrations of copper (Cu), silver (Ag), gold (Au), and palladium (Pd) that can exceed up to 40 times more than those found in their natural ores [15, 30, 36, 91]. Therefore, waste from PCBs is considered as city ores for metal recovery [70, 89, 110, 113]. Mobile phones proved to be the highest in WPCBs content (26%) among other types of e-waste as shown in Table 1 [35, 111].

Due to a variety of economic and environmental concerns, precious metal contents of WPCBs, which are the most valuable substance in mobile phones, are currently the focus of various research and industries [14, 97]. Around 15% of WPCBs go through some sort of recycling processes [87]. While the remaining are either incinerated or sent to landfill which can pose serious environmental and health issues by the production of heavy metals and toxic compounds [42, 49, 77]. Therefore, the recycling of WPCBs is necessary not only to mitigate environmental impact, but is also considered as a vital secondary source for precious materials [15, 23, 47, 49, 100]. From economic point of view, WPCBs are considered as the most attractive portion of e-waste [25, 36, 38]. For example, according to a study by Kumar et al. [53, 54], 1 million units of mobile phones (148.3 tons) can be treated to recover 9 kg of Pd, 24 kg Au, 250 kg Ag, and 9000 kg Cu, compared to 3333×10^3 kg of Pd ore, $23{,}762 \times 10^3$ kg Au ore, 1160×10^3 kg Ag ore, and 1500×10^3 kg of Cu ore, that will be needed to recover the same amount of metals from their natural mines which is very high as compared to recover them from WPCBs [53, 54].

Various techniques are employed for the recovery of metals from WPCBs including mechanical method [86], pyrolysis [13, 19], bio-metallurgical [15], and a combination of these techniques [34, 55, 66, 67], with hydrometallurgical [39, 59], 62, 75, 107 and pyrometallurgical [9, 60, 72, 73] being the two main techniques for metals recovery from WPCBs [17]. Despite the fact that the latter has been reported by several studies as it is widely used for recovering precious metals and non-ferrous from WPCBs [36], this technique causes a significant environmental damage [63], in addition to many health problems such as respiratory diseases and cancers [17]. High-temperatures heating of WPCBs generate hazardous gases, which must be removed from the air through the implementation of flue gas purification systems [88]. Furthermore, because of the high energy usage and the additional air filtration system, this method is quite expensive which makes it not very efficient for recycling WPCBs of mobile phones particularly in medium and small industries [32, 95].

Nowadays, the hydrometallurgical technique is considered to be a more efficient alternative as it starts to grab more attention compared to pyrometallurgical technique for WPCBs recycling [58]. The hydrometallurgical technique exhibited significant advantages for metal recovery from WPCBs (illustrated in Table 2) as it is more precise, predictable, easy to control, and requires comparatively low capital and operating costs and energy, highly selective extraction of metals, enabling the potential for recycling of heterogenous metals, in addition to having a lower footprint on the environment due to the negligible emission of hazardous gases [2, 28, 96, 107]. All these factors make hydrometallurgical technique advantageous over the other techniques used for the recovery of precious and base metals from WPCBs.

Generally, as presented in Fig. 1, the hydrometallurgical technique usually comprises of various steps including pre-processing then the leaching process followed by a various purification and recovery processes of WPCBs [83]. During the leaching process, inorganic, base (alkaline), and organic solutions are commonly utilised to solubilize solid metals [10, 15, 16, 64, 81, 99]. The leaching process relies on various aspects for maximum metal dissolution into solution, including the choice of the suitable leaching agent, physical and chemical properties of leached metals, selectivity, and cost of the leaching agents, as well as their environmental consequences [7].

As for the purification and recovery processes, several methods are used such as electrowinning (chemical deposition), cementation, chemical precipitation, solvent extraction, and ion exchange [62]. The hydrometallurgical approach opens new fields of study, taking into consideration global development, rising demand for technology, and natural resource depletion of many metals. As a result, we have concentrated

Table 2 Advantages of hydrometallurgical techniques over pyrometallurgical techniques

	Hydrometallurgical process	Pyrometallurgical process
Environmental impacts	Low	High
Capital and operation costs	Low	High
Hazardous and toxic gases emission	Possibility to treat or negligible	High generated
Selectivity of recovered metals	High	Low
Metals recovery	High	Low
Energy Consumption and costs	Low	High

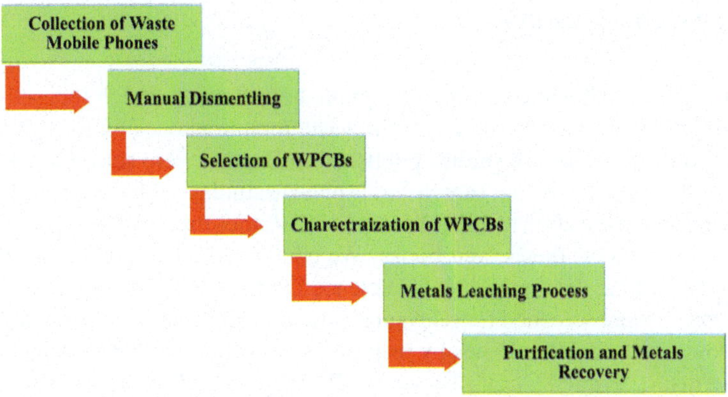

Fig. 1 Main phases of hydrometallurgical process

our attention on hydrometallurgical techniques for metals recovery from WPCBs of mobile phones.

Up to date, most of the published reviews about the recovery of metals from WPCBs are mainly focused on generation and management, environmental and social impact, policy and legislation, and metals recovery [1, 4, 11, 27, 33, 35, 37, 54, 62].

Although hydrometallurgical technique has gained attention in the past few years, only few published reviews discussed metal recovery of WPCBs from mobile phones [1, 62, 74, 112, 122]. Most of these studies focused on the hydrometallurgical technique to recover only precious metals ignoring heavy metals such as (Cr, As, Cd, Pb and Hg) as well as base metals (Fe, Al, Sn, Cu, Zn, etc.) [9, 14, 35, 62, 117, 118] that make up the majority of the metal content of WPCBs as will be discussed in detail in the following sections of this review. Up to our knowledge, no systematic review has been done on hydrometallurgical recovery of precious and base metals from WPCBs of mobile phones in particular. This systematic review provides a comprehensive up-to-date summary and analysis of the existing hydrometallurgical techniques for the recovery of precious and base metals from WPCBs particularly from mobile phones in recent years. The use of hydrometallurgical in hybrid techniques such as bio-hydrometallurgical-hydrometallurgical, pyrometallurgical-hydrometallurgical and others are excluded as they are considered beyond the scope of this review. The emphasise of this review is on the chemical composition of WPCBs, comparison of different leaching techniques for Cu, Au and other metals, and the various purification and recovery techniques for each leaching process are discussed in detail to present an insight for the selection of suitable leaching agents (inorganic, alkaline, organic) and purification techniques.

2 Research Methods

We performed a systematic search of the published research, which focused on recycling printed circuit boards (PCBs) from waste mobile phones by using hydrometallurgical process. Relevant studies published in English were selected, while other languages were excluded. Our search was limited to studies published in the last seven years, which are from 2015 to 2021. We excluded studies that reported outcomes from reviews, conferences, thesis, and books. We conducted a comprehensive literature review using online databases such as Science Direct, Web of Science, and Scopus. The databases were screened for relevant peer reviewed papers focusing on recycling PCBs from waste mobile phones by using hydrometallurgical process based on the search terms: ((“Printed Circuit Boards” OR “PCBs”) AND (“Mobile Wastes” OR “Spent Mobile” OR “Discarded Mobile”) AND (“Hydrometallurgical Process”)). The initial search in Scopus, Web of Science, and Science Direct was conducted in October 2021 and yielded 114 hits (step 1). Other additional sources included 15 papers were then assessed for another addition. Based on the exclusion criteria, 64 papers were excluded because they were not original research articles, not limited to the selected period from 2015 to 2021 and are duplicated and not relevant research, which resulted in 3 additional papers (step 2). Based on a screening of titles and abstracts, 48 papers were excluded because they did not focus on recycling PCBs from mobile wastes by using hydrometallurgical. We excluded studies reporting results in PCBs from computers, bio- hydrometallurgical processes, pyrometallurgical processes, Lithium-ion batteries from mobile wastes, reviews, conferences, short communications, thesis, books and those that did not report any outcome related to hydrometallurgical processes (step 3). This resulted in 9 papers to be further assessed (step 4). After preliminary screening, studies deemed relevant were retrieved for assessment. Eligibility was assessed independently, and disagreements were resolved by consensus, for all of the included reports. Hence, a total of 9 research papers were included. We followed the broad search strategies suggested for systematic reviews to ensure that no publications were overlooked and also assessed the cited references lists of the included articles for further relevant research [26]. For the review search, we used the Preferred Reporting Items for Systematic Reviews and Meta-Analyses (PRISMA) guidelines [71], and the process is illustrated in Fig. 2.

3 Analytical Methods of Chemical Composition of WPCBs from Mobile Phones

Various methods were utilized to determine the composition of metals obtained from WPCBs as their total composition has been analysed by many researchers using different analytical methods of chemical composition of metals from WPCBs of mobile phones as illustrated in Table 3. It was noticeable that different studies employed several methods to analyze the metals found in the WPCBs depending on

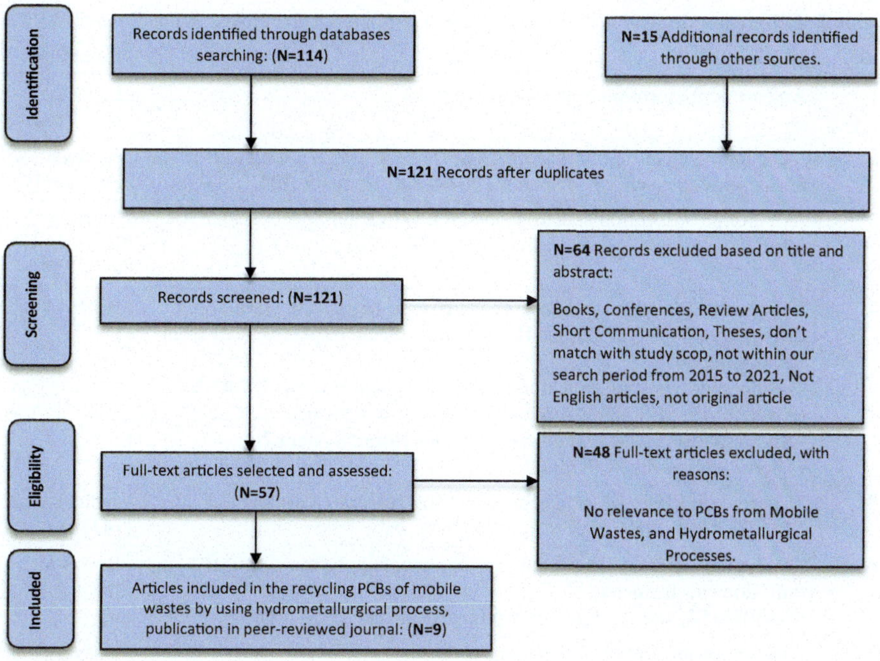

Fig. 2 Flowchart of study selection of the systematic literature review through review based on PRISMA guidelines

the availability of devices in their laboratories and the type of metals to be detected. We found that the most used method is the Atomic Adsorption Spectrometry (AAS). Silva et al. [98] was the only one to not to mention in detail the analytical methods they used in their study to determine the compositions of metals found in the tested WPCBs.

4 Content of Basic and Precious Metals in WPCBs from Mobile Phones

Although PCBs are small in size, they comprise numerous precious and base metals that account for 6% of total mass of e-waste [65]. In addition, WPCBs from mobile phones hold more precious metals than other small-sized e-waste, making them economically attractive for recycling [104]. The metals content in WPCBs are categorized into precious metals (Au, Ag, Cu) and base metals (Fe, Al, Ni, Zi, Sn, Pb, Mn) which are shown in Table 4. As can be noticed from Table 4, Cu is the most abandoned element in WPCBs as mentioned in most studies with a share of around 88 wt% [84]. In addition, Ni, Fe, Zn, Al, Au and Sn are other significant metals found within WPCBs with different share in terms of weight percentages.

Table 3 Different analytical techniques used to determine the chemical compositions of WPCBs

No.	Used methods	References
1	An atomic Adsorption Spectrometry (AAS)	Lisińska et al. [65]
2	"A scanning electron microscope (SEM) (Zeiss EVO18 with LaB6 filament) with an energy dispersive X-ray spectrometer (EDX) (Oxford)"	Barnwal and Dhawan [6]
3	"An atomic absorption spectroscopy (AAS) (Varian Spectrometer AA 20+)"	Martina Laubertová et al. [57]
4	"The technique of X-ray fluorescence spectrometry with wavelength dispersion (WDXRF; Primus II X-ray fluorescence spectrometer from Rigaku, Japan)" "Using microwave plasma atomic emission spectroscopy (MPAES) (Agilent MP-AES 4200, Agilent Technologies, Chicago, IL, USA)" "An X-ray energy dispersion spectrometer Noran System 7 (Thermo-Scientific Waltham, MA, USA)"	Brožová et al. [7]
5	"Microwave Plasma Atomic Emission Spectroscopy (Agilent MPAES-4210, Malaysia)"	Jadhao et al. [41]
6	"An Optima 8300 ICP-OES inductively coupled plasma—optical emission spectrometer" "An Elico SL 168 Atomic Absorption Spectrometer (AAS)"	Rao et al. [84]
7	"An atomic absorption spectroscopy"	Kasper et al. [50]
8	Not mentioned	Silva et al. [98]
9	"An atomic absorption spectroscopy (AAS) (AAnalyst™ 200, Perkin Elmer)"	Das et al. [17]

The results of these studies showed that there are differences in the amounts of precious and base metals found in WPCBs tested. For example, Martina Laubertová et al. [57] found that copper concentration constitutes 19.09 wt% of the WPCBs used in their study, meanwhile Das et al. [17] reported Cu percentage to be 66 wt% of the metal content of the WPCBs used in their experiments. The same observation can be made in the case of Ni whereby its concentration was found to be 1.26 wt% and 2.3 wt% in the two previous studies, respectively. This difference may be attributed to two main factors, the year of manufacturing and several types of WPCBs used which could contain different amounts of metals [31, 80], whereby their complexity contrasts by the specific functions associated with them and the quantity of those functions [50]. Varied concentrations of other metals present in tested WPCBs are illustrated in Table 4 according to data published by different authors. Consequently, these results of the analytical chemical composition showed that the WPCBs from mobile phones contain plenty of precious and base metals in their composition which makes their recycling much more attractive economically.

Table 4 Concentrations of different metals in WPCBs from mobile phones in wt%

No.	References	Cu	Au	Ag	Al	Pb	Zn	Ni	Mn	Fe	Sn	Others
1	Martina Laubertová et al. [57]	19.09	0.04	–	0.83	0.40	1.78	1.26	–	1.96	1.29	–
2	Brožová et al. [7]	21.46	0.10	0.37	2.97	-	0.34	–	–	4.67	1.49	68.59
3	Jadhao et al. [41]	24.51	–	0.07	3	0.25	1.07	0.54	0.03	0.21	0.90	–
4	Rao et al. [84]	83 ± 5	0.05 ± 0.01	–	–	–	–	2.3 ± 1	–	–	–	–
5	Kasper et al. [50]	0.142–0.7	–	–	–	–	–	–	–	–	–	–
6	Silva et al. [98]	–	–	–	–	–	–	–	–	–	–	–
7	Lisińska et al. [65]	–	–	–	–	–	–	–	–	–	–	–
8	Barnwal and Dhawan [6]	–	–	–	–	–	–	–	–	–	–	–
9	Das et al. [17]	66	0.05	–	–	–	–	2.30	–	–	–	31.66

5 Leaching Techniques

After characterization, the leaching process is an essential step in the hydrometallurgical treatment of WPCBs, whereby many researchers performed different leaching tests with various organic and inorganic acids, and bases as leaching agents under different conditions [6, 7, 17, 41, 50, 54, 65, 83, 98]. The following sections illustrate distinct types of leaching agents and the best types of oxidants used to extract Cu, Au, and other main base metals found in WPCBs. It is well known that selecting the appropriate leaching agents is critical for achieving maximal rate dissolving of the metals contained in WPCBs into the solution. For example, parameters such as the selectivity and cost of oxidant agents, the physical and chemical nature of the leached metals, and their environmental implications are important to be considered when selecting the optimal leaching agents [7].

5.1 Copper (Cu) Leaching Techniques

Copper constitutes the highest percentage of PCBs with 20–25 wt % which makes it comes first when thinking of the treatment of WPCBs, that in addition to the fact that the recovery of copper from WPCBs is much efficient than natural copper extraction process from ores which amount for only 0.5–1% [18, 109]. Given the hig5h world consumption, the extraction of copper from WPCBs is thus considered to be an economically viable. Additionally, copper is known for its potential as a metal that is soluble in very strong oxidative acids [65], hence, as illustrated in Table , 6 and 7, the first set of copper leaching experiments using various acid mixtures showed the need for an oxidating agent to facilitate effective leaching and the ability to react with metals found in WPCBs samples at the same time [119]. Therefore, conventional acids such as sulfuric acids (H_2SO_4), hydrochloric acid (HCL), nitric acid (HNO_3), ammonia (NH_3), and ammonium sulphate ($(NH_4)_2SO_4$) were not enough to leach all copper contained in WPCBs samples considering other parameters such as temperature, reaction time, and agitation speed do not improve the processing conditions. That is why, due to the limited degree of copper dissolution, experiments in many studies of leaching copper were conducted using hydrogen peroxide (H_2O_2) [6, 41, 65], ferric sulphate ($Fe_2(SO_4)_3$) [57], and ozone (O_3) [7].

5.2 Inorganic Acidic Leaching Agents

It is noticeable that leaching tests with sulfuric acid (H_2SO_4) are heavily used in most of the studies presented in Table 5, that could be attributed to the fact that H_2SO_4 is an economical leaching agent due to its availability and relatively low price [65]. Table 5 shows how H_2SO_4 was used at different concentrations with different oxidant

ozone is more effective as it is able to work at room temperature and atmospheric pressure, along with its potential application as a leaching agent at significantly low concentrations of solution. Furthermore, as mentioned earlier, Cu leaching using ozone is considered environmentally friendly due to the harmless metal content in the solution during and after the leaching process [7].

In addition, Aqua Regia is another leaching agent that is a mixture of hydrochloric acid and nitric acid ($HCL + HNO_3$) which when used to extract copper it yields up to 9.5% as reported by Das et al. [17]. On the other side, 75% is the higher percentage of Cu extraction that can be achieved using the same leaching agent, aqua regia [6]. We think that the reason behind the huge difference in Cu yield in both studies may be attributed the difference in aqua regia concentration used as we don't know the exact concentration in [17] since it wasn't mentioned in their paper compared to Barnwal and Dhawan [6] study. This illustrates the importance of selecting the right concentration of the leaching agent as it can significantly improve the yield percentage of Cu.

5.3 Base Leaching Agents

Ammonia (NH_3) is a selective leaching agent of base metals that is inexpensive and easily available. The use of ammonia which is an alkali leaching agent gained more attention compared to strong acid leaching for its selectivity [69]. However, to enhance the economic efficiency of the metal extraction process, reductants must be added into the ammonia solution [52]. Jadhao et al. [41] reported a significant recovery of Cu using ammonia and ammonium sulphate as a reductant in a leaching solution [41].

As illustrated in Table 6, the authors found 100% leaching of Cu when employing hydrogen peroxide as an oxidizing agent [41]. Das et al. [17] have investigated the recovery of Cu using ammonium thiosulphate ($(NH_4)_2S_2O_3$). The authors also found 2.6% leaching of Cu without employing any oxidizing agents [17].

5.4 Organic Acidic Leaching Agents

Despite the fact that organic acids are generally not preferred for the leaching of precious metals due to their relatively low boiling temperatures, weak acidities and decomposition. Das et al. [17] conducted leaching experiments employing various organic acids including Oxalic Acid ($C_2H_2O_4$), Gluconic Acid ($C_6H_{12}O_7$), Citric Acid ($C_6H_8O_7$) for the purpose of recovering Cu. The results showed that each organic acid used yielded a Cu percentage of 1.38%, 0.139%, 2.527% for Oxalic Acid, Gluconic Acid and Citric Acid, respectively. The low extraction recovery rates of Cu in [17] leaching tests could be attributed to the absence of oxidant agents, as

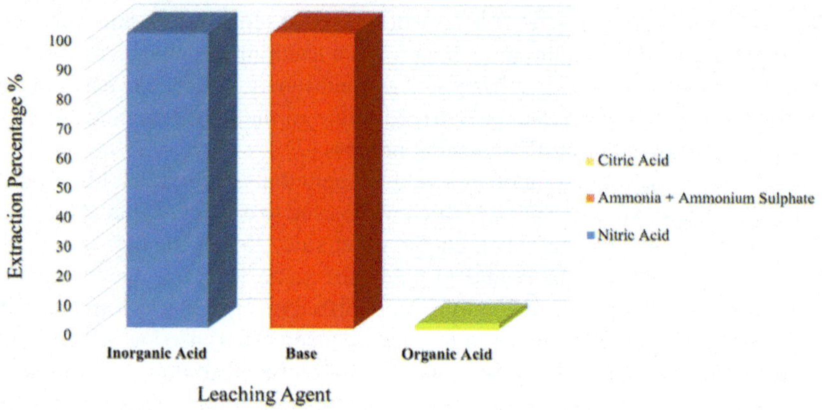

Fig. 3 Best inorganic, organic and basic leaching agents for Cu extraction

well as some missing parameters such as organic acids concentrations that were not considered in their study [17].

Overall, organic leaching agents do not show a better amount of Cu extraction rate than other inorganic and base leaching agents. Among various inorganic acids, nitric acid shows the highest Cu leaching percentage, which is 99–100% in most studies. In contrast, the low concentration of nitric acid, effective temperature and time of extraction may all be reasons for the low extraction rate 11.39% of Cu obtained by Das et al. [17]. Hydrogen peroxide is the best oxidising agent to be used with sulphuric acid where it can lead to a Cu extraction rate of up to 100% similarly, the presence of strong oxidant like ozone significantly increases Cu extraction rate to around 70% when used with hydrochloric acid, in contrast to the low Cu extraction rate (7.82%) obtained in the absence of an oxidising agent. Furthermore, aqua regia shows the highest Cu extraction rate at 75% in the presence of hydrogen peroxide. As for the base leaching agent, ammonia + ammonium sulphate shows the highest extraction rate of Cu with 100% at the present of hydrogen peroxide as oxidising agent, meanwhile, ammonium thiosulphate shows the lowest Cu extraction of 2.6%. When it comes to organic leaching agents, citric acid shows the highest Cu extraction rate of 2.527%. Hence, comparing all the leaching agents, Fig. 3 illustrates the best inorganic, organic and base leaching agents for Cu extraction.

5.5 *Gold (Au) Leaching Techniques*

Gold (Au) is the most precious metal found in WPCBs due to its intrinsic value in addition to its abondance, where its concentration is 35–50 times higher than that of its natural ore [79]. Although, as shown in Table 4, the concentration of Au is the lowest among all metals found in WPCBs leading to many challenges for its recovery

[22], mobile phone recycling is still rewarding as it could contain up to 0.34 kg of Au per ton [78]. Table 8 shows the different leaching rates of gold using different leaching agents.

In characterization tests, amounts of Au based on the finding obtained for each type of WPCB samples showed that each type of WPCBs contain diverse amount of Au based on many parameters like year of manufacturing, complexity and the quantity of functions associated with those devices [31, 80]. Table 4 displays the various gold concentrations found in the literature.

After performing characterization, Kasper et al. [50] used different leaching agents such as cyanide-based, sodium thiosulphate ($Na_2S_2O_3$) and ammonium thiosulphate ($(NH_4)_2S_2O_3$) under various conditions to extract Au from their samples. Commercial stripping using cyanide is a conventional method that has been in use for over a century for the extraction of gold from its primary ores, and it is being used to extract gold from secondary sources. Aside from the numerous advantages of using cyanide such as the simplicity of its process, high efficiency, and relatively low cost, cyanide has a full set of drawbacks including high degree of toxicity for both humans and the environment as it has high capacity to cause contamination in addition to the need for precise control in effluent treatment. Au leaching using cyanide was carried out in the following conditions: temperature 296.15 $\pm$ 300.15 K, 2 h of extraction time, and with solid: liquid ratio 1:25 in a closed reactor. This resulted in an Au yield of 88% which is the highest extraction rate compared to other leaching agents that will be explained in the coming sections.

Although Au extraction using cyanide as a leaching agent is effective and economical, but it is still not environmentally friendly due to its high toxicity [50]. Thiosulphate has been advocated as an alternative to cyanide since it might be applied for gold ores that cannot be extracted by the conventional cyanidation method, in addition to its low toxicity, relatively low cost [94, 121], and faster leaching rate [12, 45]. One main downside of using thiosulphate as a leaching agent is the extremely slow leaching reaction rate. Therefore, copper ions and ammonium are added to accelerate the chemical reaction rate [50]. Ammonium thiosulphate and sodium thiosulphate are two leaching agents that were used by Kasper et al. [50] that yielded 75% and 70% of Au respectively at the same following conditions: leaching agent concentration of 0.12 M, copper sulphate ($CuSO_4$) 20 mM, temperature 301.15 $\pm$ 305.15 K, 4 h of extraction time, and with solid:liquid ratio 1:25 in a closed reactor. Although ammonium thiosulphate showed a higher extraction rate of Au compared to sodium thiosulphate, but the latter is more preferable due to its cost (approximately eight times cheaper than ammonium thiosulphate). Additionally, Das et al. [17] obtained an extraction rate 69.4% of Au using ammonium thiosulphate, which proofs the optimizing effect of copper sulphate in obtaining higher extraction rate as in the case of [50].

Although nitric acid (HNO_3) was shown to offer a superior leaching performance in the case of Cu extraction with an extraction rate of 99.9%, nitric acid did not show the same performance in the case of Au despite using typically optimal conditions, 3 M (HNO_3), temperature 30 °C, extraction time 2 h, solid:liquid 50, nitric acid leached almost no Au (Au = 0%) into the solution but this process of selective

Table 8 The different amounts of gold (Au) found in the literature

	Leaching agents	Concentration of leaching agents	Oxidant agents	Effective temperature (K)	Effective time (h)	Solid/liquid	Recovered metals	References
Inorganic acids	Commercial stripping			296.15 ± 300.15	2	1/25	Au = 88%	Kasper et al. [50]
	Hydrochlroic acid (HCl)	2 M	Hydrogen peroxide (H_2O_2) = 30%	353	4	1:4	Au 12.46 g/dm^3	Brožová et al. [7]
		2 M	Ozone (O_3) = 140 mg/dm^3	353	2	1:4	Au = 6.89 g/dm^3	Brožová et al. [7]
	Nitric acid (HNO_3)	3 M		303.15	2	50	(0.05 ± 0.01) wt% ~ 100%	Rao et al. [84]
	Ammonium thiosulphate ($(NH_4)S_2O_3$)	0.12 M	Copper sulphate ($CuSO_4$) = 20 mM	301.15 ± 305.15	4	1/25	Au = 75%	Kasper et al. [50]
							Au = 0.031 wt% ~ 69.4%	Das et al. [17]
	Sodium thiousulphate ($Na_2S_2O_3$)	0.12 M	Copper sulphate ($CuSO_4$) = 20 mM	301.15 ± 305.15	4	1/25	Au = 70%	Kasper et al. [50]
	Aqua regia (hydrochloric acid (HCl) + nitric acid (HNO_3))					1/10	Au = 0.04 wt% ~ 88.8%	Das et al. [17]
	Urea (CH_4N_2O)	0.2–0.5 M/100 m	Sodium bisulphate ($Na_2S_2O_3$) = 0.001–0.02 M			1/25	Au = 4.1 g	Barnwal and Dhwan [6]

dissolution of the base metals is often required as the first step to concentrate Au into a solid residue. The concentrated Au found in the solid form among the separated residue which was afterwards dissolved in aqua regia and analysed by ASS, whereby the main metal of the solid residue was Au = 100% [84].

Hydrochloric acid (HCL) is regarded as an ineffective leaching agent when used for Au extraction without the addition of an oxidising agents. Therefore, there is a need for the addition of oxidising agents to enhance its leaching capability. In Brožová et al. [7] experiment, 30% hydrogen peroxide was added to 2 M of HCl at 353 K with a solid:liquid ratio of 1:4 for 4 h, yielding 12.46 g/dm^3 Au. On the other hand, 6.89 g/dm^3 of Au was obtained upon the addition of ozone (140 mg/dm^3) as an oxidant agent, to 2 M of HCL at 353 K for 2 h with a solid:liquid ratio of 1:4. When compared to hydrogen peroxide, ozone is regarded more environmentally friendly due to the harmless metal presence in the solution during and after the leaching process [7].

Aqua regia is a known strong leaching agent that is commonly used to precipitate Au from a leach solution. Nevertheless, aqua regia is not widely used in full scale operations due to high corrosion to other equipment [29, 46]. Das et al. [17] and Rao et al. [84] acquired 88.88% and 100% of Au respectively, proving that all of the Au found in the solid residues of their experiments have been extracted by aqua regia [17, 84].

Barnwal and Dhawan [6] used urea (CH_4N_2O) as leaching agent to neutralize the leach solution for Au dissolution, followed by sodium bisulphite ($Na_2S_2O_5$) which used for precipitation of Au in their experiment. 4.1 g of Au was obtained by treating 10 kg of WPCBs.

Overall, although aqua regia as well as cyanide showed the highest percentage of Au extraction at 100% and 88% respectively at all the previous studies, these two leaching agents are not preferred to be used due to several negative impacts on the environment and equipment involved. From environmental point of view, ammonium thiosulphate and sodium thiosulphate showed the best leaching capability upon the addition of copper sulphate at 75% and 70% respectively compared to the two previously mentioned leaching agents.

5.6 *Other Metals Leaching Techniques*

The recovering of valuable base metals other than Cu such as Tn, Pb, Zn, and Ni is of a great significance from environmental and economic point of views which made them attractive to be extracted. Hydrochloric acid has caught the attention of many researchers recently [7, 17]. Whereby different studies used different inorganic leaching and oxidizing agents to extract various metals as demonstrated in Table 9. Brožová et al. [7] used 2 M hydrochloric (HCL) at the presence of 140 mg/dm^3 of ozone at 353 K for 4 h which resulted in leaching Ag (12.94 g/dm^3) and Tin (100 g/dm^3). On the other hand, when replacing ozone with 30% of hydrogen peroxide under the same conditions, only Tin was extracted at same quantity. In the absence of an oxidant agent, Das et al. [17] showed that hydrochloric acid has the ability to extract various metals such as Pb (3.66 wt%), Fe (7.18 wt%), and Zi (0.74 wt%).

A mixture of hydrogen fluoride and hydrogen peroxide ($HF + H_2O_2$) showed the highest percentage of extracting of zinc and nickel with 99.5 wt% for each at a concentration of 5 mol/L each [98]. On the other hand, Das et al. [17] extracted Ni, Fe, and Zn at rates of 0.34, 10.74, and 1.66 wt% respectively. It is noticeable that 2.30 wt% of Ni was detected in their samples as displayed in Table 4. But after the leaching process, low rate (0.34 wt%) of Ni was obtained. These results show the importance of adding an oxidant to increase the extraction rates. Lisińska et al. [65] used 5 M sulphuric acid with 10% of hydrogen peroxide as an oxidizing agent at 343 K for 72 h which resulted in extracting Fe (3.87–6.36 g/dm^3) and Zi (0.1–0.47 g/dm^3). However, from economical point of view, it has low efficiency due to its high cost and the fact that it is not really effective in increasing the leaching level of other metals [65].

As for nitric acid, much more metals were extracted compared hydrochloric and sulphuric acid. Rao et al. [84] results confirmed that Ni, Zn, Pb, Cd, Al, Pd, and Sn, Ag were extracted using 3 M nitric acid at 303.15 K for 2 h which resulted at extraction rate of more than 99% for all the mentioned metals except for Sn (30.8%) and Ag (8%) [84]. Das et al. [17] used nitric acid as well in their study where the result showed the extraction of Pb (4.94 wt%), Fe (12.76 wt%), Zn (2.04 wt%), and Ni (0.66 wt%) [17].

As mentioned earlier, Ammonia (NH_3) as an alkali leaching agent, has attracted more attention than strong acid leaching. Jadhao et al. [41] reported a high recovery of Ni and Zn using ammonia and ammonium sulphate as a reductant in a leaching solution at the presence of 0.4 M hydrogen peroxide (H_2O_2) which yielded 90% and 75% of Ni and Zn respectively [41]. On the other hand, Das et al. [17] used a different alkaline leaching agent, ammonium thiosulphate, ($(NH_4)_2S_2O_3$). The results showed that several metals were leached with different wt% of Pb = 0.006375 wt%, Ni = 0.035125 wt%, Zn = 0.00575 wt% [17], as presented in Table 10. This illustrates the high selectivity of this alkaline leaching agent. However, the low outputs of the leached metals could be attributed to the absence of an oxidant agent.

As already stated above, organic acids are not recommended for precious metal leaching due to their weak acidity. Oxalic Acid $C_2H_2O_4$, Gluconic Acid $C_6H_{12}O_7$), Citric Acid ($C_6H_8O_7$) are all organic leaching agents that are used for the recovery of different metals. The results showed that different wt% percentage were obtained at same temperature (353.15 K) and leaching time (4–5 h) for the 3 organic leaching agents without mentioning their concentrations. Citric Acid showed the highest leaching rate for Pb, Ni, Fe, and Zn followed by Oxalic Acid and Gluconic Acid. 1.0105 wt%, 0.7306 wt%, 0.09525 wt%, 0.289 wt% of Pb, Ni, Fe and Zn respectively were leached by Citric Acid ($C_6H_8O_7$). Table 11 shows the percentage of leaching rates of some metals.

Overall, organic leaching agents do not show a better metal extraction compared to other inorganic and base leaching agents. Among various inorganic acids, nitric acid and Hydrogen Fluoride and Hydrogen Peroxide showed the highest Pb, Ni, Zn and Fe leaching rates, which are more than 99% with Nitric Acid having the highest

Table 9 Other metals extraction using inorganic acids

	Leaching agents	Concentration of leaching agents (M)	Oxidant agents	Effective temperature (K)	Effective time (h)	Recovered metals	References
Inorganic acids	Sulfuric acid (H_2SO_4)	5	Hydrogen peroxide (H_2O_2) = 10%	343	72	Fe = 6.36–3.87 g/dm^3 Zn = 0.47–0.1 g/dm^3 Al = 0.09–0.42 g/dm^3	Lisińska et al. [65]
		Not specify	Not specify	–	–	Ni = 0.34 wt% Fe = 10.74 wt% Zn = 1.66 wt%	Das et al. [17]
	Hydrochloric acid (HCl)	2	Ozone (O_3) = 140 mg/dm^3	353	4	Ag = 12.94 g/dm^3	Brožová et al. [7]
		2	Ozone (O_3) = 140 mg/dm^3	313 and 353	4	Sn = 100 g/dm^3	Brožová et al. [7]
		2	Hydrogen peroxide (H_2O_2) = 30%	298, 313 and 353	4	Sn = 100 g/dm3	Brožová et al. [7]
						Pb = 3.66 wt% Fe = 7.18 wt% Zn = 0.74 wt%	Das et al. [17]

(continued)

Table 9 (continued)

Leaching agents	Concentration of leaching agents (M)	Oxidant agents	Effective temperature (K)	Effective time (h)	Recovered metals	References
Nitric acid (HNO_3)	3		303.15	2	Ag = 8% Pb = 99.3% Ni = 99.9% Zn = 99.5% Al = 99.3% Cd = 99.9% Pd = 99.9% Sn = 30.8%	Rao et al. [84]
					Pb = 4.94 wt% Fe = 12.76 wt%	Das et al. [17]
Hydrogen fluoride and hydrogen peroxide ($HF + H_2O_2$) = 5 mol/L and 5 mol/L			303.15	1	Zn = 99.5 wt% Ni = 99.5 wt%	Silva et al. [98]

Table 10 Other metals extraction using alkaline

	Leaching agents	Concentration of leaching agents	Oxidant agents	Effective temperature	Effective time	Recovered metals	References
Base	(1) Ammonia NH_3^+ (2) Ammonium sulphate $(NH_4)_2SO_4$	(1) 90 g/L (2) 180 g/L	Hydrogen Peroxide (H_2O_2) = 0.4 M	353.15 K	4 h	Ni = 90% Zn = 75%	Jadhao et al. [41]
	Ammonium thiosulphate $((NH_4)_2S_2O_3)$	–	–	–	–	Pb = 0.01 wt% Ni = 0.02 wt% Zn = 0.01 wt%	Das et al. [17]

Table 11 Other metals extraction using organic acids

	Leaching agents	Concentration of leaching agents	Oxidant agents	Effective temperature (K)	Effective time (h)	Recovered metals	References
Organic acids	Oxalic acid $(C_2H_2O_4)$	–	–	353.15	4–5	Pb = 0.34 wt%, Ni = 0.17 wt%, Fe = 0.31 wt%, Zn = 0.05 wt%,	Das et al. [17]
	Gluconic acid $(C_6H_{12}O_7)$	–	–	353.15	4–5	Pb = 0.17 wt%, Ni = 0.02 wt%, Fe = 0.01 wt%, Zn = 0.01 wt%	
	Citric acid $(C_6H_8O_7)$	–	–	353.15	4–5	Pb = 1.01 wt%, Ni = 0.73 wt%, Fe = 0.095 wt%, Zn = 0.29 wt%	

selectivity as it leached more metals such as Al, Cd, Pd Sn and Ag. In contrast, the absence of an oxidant agents in the case of organic leaching agent might contribute to the low leaching rate of different metals. This requires further research to be done on the effects of the concentration of the organic leaching agent and the addition of oxidant agent on the leaching rate.

6 Purification and Recovery

After leaching process, metal recovery is the second crucial step in hydrometallurgical operations. Various methods such as electrowinning (chemical deposition), cementation, chemical precipitation, solvent extraction, and ion exchange are exploited for the purification of base and precious metals based on the specific leaching solutions [62]. Hence, the following section discusses the various individual and hybrid purification methods used for Cu and Au recovery as shown in Table 12.

Table 12 Different Au and Cu purification methods from the literature

Leaching agents	Leached solution	Target recovered metals	Purification methods	Purity of recovered metals	References
Inorganic leaching agent	Thiosulfate	Au and Cu	Electrowinning	Au = 94% and Cu = 95%	Kasper et al. [50]
	Aqua regia	Cu	Electrowinning	Cu = 97%	Barnwal and Dhawan [6]
	Sulfuric acid and hydrochloric acid solutions with O_3 in addition	Cu	Cementation and electrolysis	24% copper residue in HCl + O_3 solution Cu was selectively cemented After electrolysis—64% copper residue in $H_2SO_4 + O_3$	Brožová et al. [7]
	Hydrogen fluoride and hydrogen peroxide (HF + H_2O_2)	Cu, Ni and Zn	Solvent extraction (SX) and chemical precipitation	99.5% of the Cu, Ni and Zn	Silva et al. [98]
Alkali leaching agent	Ammoniacal	Cu	Electrowinning	Cu = 98.38%	Jadhao et al. [41]

Leaching reagent systems, metal concentration in the leaching solution, the impurities present, and the category of target metals to be recovered all play an essential role in determining the metal recovery and the selection of downstream purification methods [21, 62, 107]. All these factors along with the leaching process contribute into recovering the final metal products. Therefore, many studies used various methods to transfer the metals in WPCBs into solution.

Due to the high concentration of Cu in WPCBs as proved in Table 4, several studies put massive attention into the recovery of Cu. In addition, although Au is usually found in low concentrations in WPCBs, it gained as much attention due to its economic value. For instance, Kasper et al. [50] used selective electrowinning by controlling electrochemical parameters such as productivity, electrical efficiency and energy consumption. Using electrode potentials of − 500 and − 700 mV versus Ag/AgCl, about 94% of Au and 95% of Cu proved to be selectively deposited from an ammoniacal thiosulphate leaching solution of WPCBs. According to the findings, the current efficiency of gold in the experiments in the real solutions was as low as 2.2% [50], which could be attributed to the fact that the concentration of copper (1130 ppm) is much greater than the concentration of gold (45.1 ppm) and the presence of other metals in the solution. It has also been observed that the productivity decreases with time as the current density decreases, which is proportional to the concentration, leading to a low speed of gold recovery over time. The energy consumption increases overtime, as the concentration of gold ions to be extracted from the solution decreases. Overall, the findings obtained from their study on the efficiency of the electrowinning process might be quite satisfactory.

In addition, Barnwal and Dhawan [6], precipitated Au from a leaching solution using Aqua regia-sodium bisulfite system, and then to neutralize the leach solution, Urea (CH_4N_2O) was exploited. It can be concluded that 4.1 g of Au and 9 kg of Cu with a purity of approximately 97% can be recovered from 10 kg of WPCB treated sample using electrowinning whereby the copper was deposited on the working electrode at room temperature at ~ 83 min at a current density of − 30 mA/cm^2 [6].

While Jadhao et al. [41] investigated the selective recovery of Cu from the ammoniacal leached solution of WPCBs (with Ni and Zn previously removed), by employing electrowinning whereby a single compartment electrolysis cell was used, with a 3 × 3 cm Al cathode and Pt anode of the same size. The results showed that Cu was recovered at an Al cathode with a constant 2.0 V DC voltage and 98.38% purity [41].

On the other side, Brožová et al. [7] used a cementing column and electrolyser in the initial tests to precipitate copper from the WPCB leaching solution using 2 M hydrochloric acid and 2 M sulfuric acid solutions with O_3. Large amount of Cu was selectively cemented (3 dm^3) for (S/L = 1:4) for 2 h at room temperature. The results indicated that 24% of copper residue was detected in the HCl + O_3 solution. After that, a 1 dm^3 solution was used for one h at 25 °C, 1.5 A intensity, and 1.4 V system voltage in an electrolysis test. The result after electrolysis showed the presence of 64% copper residue in H_2SO_4 + O_3. It can be concluded that the complete recovery of Cu cannot be achieved in their experiment due to the presence of individual metals such as Ag, Au, and Sn [7]. Further research needs to be continued on ozone leaching to understand this process to select suitable parameters for effective recovery of metals from WPCB.

Due to the presence of additional metals in the leaching solution such as Fe, Solvent Extraction is an effective method for coper separation from leaching solution [51, 115]. Based on that, Silva et al. [98] recovered Cu, Zn, and Ni from Hydrogen Fluoride and Hydrogen Peroxide ($HF + H_2O_2$) leaching solution using a combination of solvent extraction (SX) and chemical precipitation techniques. About 99.5% of the Cu, Ni and Zn were precipitated at 303.15 K, for 1 h. Recovery of fluoride ions from the final products was another target for both environmental and economic reasons [98].

In summary, from environmental and economic point of view, electrowinning has vast superiority in comparison to other purification methods as it requires simple electric devices with minimal chemical input. Nevertheless, its low selectivity makes it almost not possible to purify specific metals from a leaching solution containing various metals especially from WPCBs [61]. On the other hand, solvent extraction has a good selectivity and requires mild reaction conditions in addition to its relatively easy scale up process. However, solvent extraction requires a reverse extraction at the end of the process [49, 56, 90, 92, 98]. As for the cementation method, the excess addition of cementing metal will reduce the grade of target metal. Chemical precipitation is the most cost effective, simple, and current method of metal separation in hydrometallurgical routes [8, 20, 68, 93]. However, metals in small concentrations in the leaching solutions might influence the low leaching selectivity and the purity of the final products for the whole purification methods [62]. In addition, the low leaching selectivity associated with acid leaching is a major problem due to the fact that most of the metals are leached into the solution [116]. Consequently, multi-step separation process is required to recover individual metals from leaching solution for high recovery purity of target metals. Therefore, it's crucial to select a leaching agent that is efficient and has a high degree of selective recovery for precious and base metals.

7 Conclusion and Future Recommendations

WPCBs of mobile phones contain a variety of valuable metals such as copper, gold and many more. The recycling of these WPCBs for the recovery of those metals is extremely rewarding from economic and environmental perspectives. The hydrometallurgical process was investigated in terms of its advantages, such as low environmental impact and capital cost. The current review analyses the present status of metals recovery from WPCBs of mobile phones through the hydrometallurgical process with focusing on leaching and purification phases and gives the discussions regarding various proposed leaching and purification techniques. The general findings derived from this review are as follows:

- WPCBs of mobile phones contain higher concentrations of precious and base metals such as (Au, Cu, Ni, Fe, etc.) compared to other e-wastes from small devices which make them economically attractive to be recycled. The percentage of these metals varies from one WPCB to another based on the manufacturing year, their specific functions associated with them and the quantity of those functions. Future studies should focus on the recovery of metals that were not considered in the literature. Samples' manufacturing year and weight are critical parameters that should be mentioned in any future work as they are often neglected by most of the literature.
- Most of the literature about the hydrometallurgical route uses similar initial sample preparation steps but they differ in the subsequent leaching and purification processes, which encouraged us to concentrate our work on these two phases. Selecting an appropriate leaching agent relies on a number of factors, most important of which are the physical and chemical nature of the leached metals, cost, efficiency, and their impact on the equipment involved, humans, and the environment. Most studies approved that the addition of an oxidant agent significantly improves metal leaching rates as long as the oxidants have high selectivity and are able to leach several metals with less economic and environmental impact.
- Although inorganic and base leaching agents proved to be more efficient in the recovery of most of the metals found in WPCBs in high amounts compared to organic leaching agents, more studies should be done on organic leaching agents due to their enormous environmental and economic benefits with taking green chemistry aspect into consideration. Furthermore, more parameters should be studied in the case of using organic leaching agents such as the concentration of the leaching agent, effective temperature and time, and the possibility of adding an oxidant agent to enhance metal leaching rates.
- Coming to the purification phase of the hydrometallurgical route, more research should be done by using hybrid purification methods for the recovery of several metals from WPCBs of mobile phones. The selection of the purification method depends on leaching agent systems, metal concentration in the leaching solution, the targeted metals to be recovered and the cost of the purification method used. Therefore, a multi-step separation process is required to recover individual metals from leaching solution for high recovery purity of the targeted metals. Hence, it's important to select an effective leaching agent with the high selective recovery of precious and base metals.

References

1. Akcil A et al (2015) Precious metal recovery from waste printed circuit boards using cyanide and non-cyanide lixiviants—a review. Waste management (New York, N.Y.) 45:258–271. https://doi.org/10.1016/j.wasman.2015.01.017

2. Akcil A, Agcasulu I, Swain B (2019) Valorization of waste LCD and recovery of critical raw material for circular economy: a review. Resourc Conserv Recycl 149(November 2018):622–637. https://doi.org/10.1016/j.resconrec.2019.06.031
3. Arshadi M, Mousavi SM, Rasoulnia P (2016) Enhancement of simultaneous gold and copper recovery from discarded mobile phone PCBs using *Bacillus megaterium*: RSM based optimization of effective factors and evaluation of their interactions. Waste Manag 57:158–167. https://doi.org/10.1016/j.wasman.2016.05.012
4. Awasthi AK, Zeng X, Li J (2016) Integrated bioleaching of copper metal from waste printed circuit board—a comprehensive review of approaches and challenges. Environ Sci Pollution Res 23(21):21141–21156. https://doi.org/10.1007/s11356-016-7529-9
5. Bai H, Wang J, Zeng AZ (2018) Exploring Chinese consumers' attitude and behavior toward smartphone recycling. J Clean Prod 188. https://doi.org/10.1016/j.jclepro.2018.03.253
6. Barnwal A, Dhawan N (2020) Recycling of discarded mobile printed circuit boards for extraction of gold and copper. Sustain Mater Technol 25:e00164. https://doi.org/10.1016/j.susmat.2020.e00164
7. Brožová S et al (2021) Hydrometallurgical recycling process for mobile phone printed circuit boards using ozone. Metals 11(5):0–9. https://doi.org/10.3390/met11050820
8. Buzatu T et al (2013) Study concerning the recovery of zinc and manganese from spent batteries by hydrometallurgical processes. Waste Manage 33(3):699–705. https://doi.org/10.1016/j.wasman.2012.10.005
9. Cayumil R et al (2014) Generation of copper rich metallic phases from waste printed circuit boards. Waste Manage 34(10):1783–1792. https://doi.org/10.1016/j.wasman.2014.05.004
10. Chauhan G, Pant KK, Nigam KDP (2015) Chelation technology: a promising green approach for resource management and waste minimization. Environ Sci Processes Impacts 17(1):12–40. https://doi.org/10.1039/c4em00559g
11. Chen M et al (2016) Evolution of electronic waste toxicity: trends in innovation and regulation. Environ Int 89–90:147–154. https://doi.org/10.1016/j.envint.2016.01.022
12. Chen X, Zhou T (2014) Hydrometallurgical process for the recovery of metal values from spent lithium-ion batteries in citric acid media. Waste Manage Res 32(11):1083–1093. https://doi.org/10.1177/0734242X14557380
13. Chiang HL et al (2007) Pyrolysis characteristics of integrated circuit boards at various particle sizes and temperatures. J Hazard Mater 149(1):151–159. https://doi.org/10.1016/j.jhazmat.2007.03.064
14. Cui H, Anderson CG (2016) Literature review of hydrometallurgical recycling of printed circuit boards (PCBs). J Adv Chem Eng 6(1):1–11. https://doi.org/10.4172/2090-4568.1000142
15. Cui J, Zhang L (2008) Metallurgical recovery of metals from electronic waste: a review. J Hazard Mater 158(2–3):228–256. https://doi.org/10.1016/j.jhazmat.2008.02.001
16. Das A, Vidyadhar A, Mehrotra SP (2009) A novel flowsheet for the recovery of metal values from waste printed circuit boards. Resour Conserv Recycl 53(8):464–469. https://doi.org/10.1016/j.resconrec.2009.03.008
17. Das D, Mukherjee S, Chaudhuri MG (2021) Studies on leaching characteristics of electronic waste for metal recovery using inorganic and organic acids and base. Waste Manage Res 39(2):242–249. https://doi.org/10.1177/0734242X20931929
18. Desmarais M et al (2020) Biohydrometallurgical processes for the recovery of precious and base metals from waste electrical and electronic equipments: current trends and perspectives. Bioresource Technol Rep 11(May):100526. https://doi.org/10.1016/j.biteb.2020.100526
19. Evangelopoulos P et al (2019) Reduction of brominated flame retardants (BFRs) in plastics from waste electrical and electronic equipment (WEEE) by solvent extraction and the influence on their thermal decomposition. Waste Manage 94:165–171. https://doi.org/10.1016/j.wasman.2018.06.018
20. Ferella F, De Michelis I, Vegliò F (2008) Process for the recycling of alkaline and zinc-carbon spent batteries. J Power Sources 183(2):805–811. https://doi.org/10.1016/j.jpowsour.2008.05.043

21. Flandinet L et al (2012) Metals recovering from waste printed circuit boards (WPCBs) using molten salts. J Hazard Mater 213–214:485–490. https://doi.org/10.1016/j.jhazmat.2012.02.037
22. Fogarasi S et al (2014) Copper recovery and gold enrichment from waste printed circuit boards by mediated electrochemical oxidation. J Hazard Mater 273:215–221. https://doi.org/10.1016/j.jhazmat.2014.03.043
23. Fornalczyk A et al (2013) E-waste as a source of valuable metals. Arch Mater Sci Eng 63(2):87–92
24. Forti V et al (2020) The global E-waste monitor 2020: quantities, flows, and the circular economy potential. United Nations University (UNU)/United Nations Institute for Training and Research (UNITAR)—co-hosted SCYCLE Programme, International Telecommunication Union (ITU) & International Solid Waste Association (ISWA), Bonn/Geneva/Rotterdam
25. Ghosh B et al (2015) Waste printed circuit boards recycling: an extensive assessment of current status. J Clean Prod 94:5–19. https://doi.org/10.1016/j.jclepro.2015.02.024
26. Grant K et al (2013) Health consequences of exposure to e-waste: a systematic review. The Lancet Global Health 1(6):e350–e361. https://doi.org/10.1016/S2214-109X(13)70101-3
27. Guo J, Guo J, Xu Z (2009) Recycling of non-metallic fractions from waste printed circuit boards: a review. J Hazard Mater 168(2–3):567–590. https://doi.org/10.1016/j.jhazmat.2009.02.104
28. Guo Q et al (2010) Pyrolysis of scrap printed circuit board plastic particles in a fluidized bed. Powder Technol 198(3):422–428. https://doi.org/10.1016/j.powtec.2009.12.011
29. Gurung M et al (2013) Recovery of gold and silver from spent mobile phones by means of acidothiourea leaching followed by adsorption using biosorbent prepared from persimmon tannin. Hydrometallurgy 133:84–93. https://doi.org/10.1016/j.hydromet.2012.12.003
30. Hadi P et al (2015) Waste printed circuit board recycling techniques and product utilization. J Hazard Mater 283:234–243. https://doi.org/10.1016/j.jhazmat.2014.09.032
31. Hagelüken C, Corti CW (2010) Recycling of gold from electronics: cost-effective use through "design for recycling". Gold Bull 43(3):209–220. https://doi.org/10.1007/BF03214988
32. Hall WJ, Williams PT (2007) Separation and recovery of materials from scrap printed circuit boards. Resour Conserv Recycl 51(3):691–709. https://doi.org/10.1016/j.resconrec.2006.11.010
33. Hao J et al (2020) Metal recovery from waste printed circuit boards: a review for current status and perspectives. Resour Conserv Recycl 157(March):104787. https://doi.org/10.1016/j.resconrec.2020.104787
34. Havlik T et al (2010) Leaching of copper and tin from used printed circuit boards after thermal treatment. J Hazard Mater 183(1–3):866–873. https://doi.org/10.1016/j.jhazmat.2010.07.107
35. He J, Duan C (2017) Recovery of metallic concentrations from waste printed circuit boards via reverse floatation. Waste Manage 60:618–628. https://doi.org/10.1016/j.wasman.2016.11.019
36. Huang J et al (2014) Leaching behavior of copper from waste printed circuit boards with Brønsted acidic ionic liquid. Waste Manage 34(2), pp. 483–488. https://doi.org/10.1016/j.wasman.2013.10.027.
37. Huang K, Guo J, Xu Z (2009) Recycling of waste printed circuit boards: a review of current technologies and treatment status in China. J Hazard Mater 164(2–3):399–408. https://doi.org/10.1016/j.jhazmat.2008.08.051
38. Ikhlayel M (2017) Environmental impacts and benefits of state-of-the-art technologies for E-waste management. Waste Manage [Preprint]. https://doi.org/10.1016/j.wasman.2017.06.038
39. Islam A et al (2020) Advances in sustainable approaches to recover metals from e-waste—a review. J Clean Prod 244:118815. https://doi.org/10.1016/j.jclepro.2019.118815
40. Jadhao PR et al (2020) Environmentally friendly approach for the recovery of metallic fraction from waste printed circuit boards using pyrolysis and ultrasonication. Waste Manage 118:150–160. https://doi.org/10.1016/j.wasman.2020.08.028
41. Jadhao PR et al (2021) Efficient recovery of Cu and Ni from WPCB via alkali leaching approach J Environ Manage 296(June):113154. https://doi.org/10.1016/j.jenvman.2021.113154

42. Jadhav U, Hocheng H (2015) Hydrometallurgical recovery of metals from large printed circuit board pieces. Sci Rep 5(101):1–10. https://doi.org/10.1038/srep14574
43. Jeon S et al (2019) A physical separation scheme to improve ammonium thiosulfate leaching of gold by separation of base metals in crushed mobile phones. Miner Eng 138(May):168–177. https://doi.org/10.1016/j.mineng.2019.04.025
44. Jeon S et al (2020) Ammonium thiosulfate extraction of gold from printed circuit boards (PCBs) of end-of-life mobile phones and its recovery from pregnant leach solution by cementation. Hydrometallurgy 191(October 2019):105214. https://doi.org/10.1016/j.hydromet.2019.105214
45. Jha MK et al (2012) Leaching of lead from solder material of waste printed circuit boards (PCBs). Hydrometallurgy 121–124:28–34. https://doi.org/10.1016/j.hydromet.2012.04.010
46. Jing-ying L, Xiu-li X, Wen-quan L (2012) Thiourea leaching gold and silver from the printed circuit boards of waste mobile phones. Waste Manage 32(6):1209–1212. https://doi.org/10.1016/j.wasman.2012.01.026
47. Kang HY, Schoenung JM (2005) Electronic waste recycling: a review of U.S. infrastructure and technology options. Resour Conserv Recycl [Preprint]. https://doi.org/10.1016/j.resconrec.2005.06.001
48. Karwowska E et al (2014) Bioleaching of metals from printed circuit boards supported with surfactant-producing bacteria. J Hazard Mater 264:203–210. https://doi.org/10.1016/j.jhazmat.2013.11.018
49. Kasper AC et al (2011) Printed wiring boards for mobile phones: characterization and recycling of copper. Waste Manage 31(12):2536–2545. https://doi.org/10.1016/j.wasman.2011.08.013
50. Kasper AC et al (2018) Electrochemical study of gold recovery from ammoniacal thiosulfate, simulating the PCBs leaching of mobile phones. Electrochimica Acta 259:500–509. https://doi.org/10.1016/j.electacta.2017.10.161
51. Kavousi M et al (2017) Selective separation of copper over solder alloy from waste printed circuit boards leach solution. Waste Manage 60:636–642. https://doi.org/10.1016/j.wasman.2016.07.042
52. Ku H et al (2016) Recycling of spent lithium-ion battery cathode materials by ammoniacal leaching. J Hazard Mater 313(March):138–146. https://doi.org/10.1016/j.jhazmat.2016.03.062
53. Kumar A, Holuszko M, Espinosa DCR (2017) E-waste: an overview on generation, collection, legislation and recycling practices. Resour Conserv Recycl, 32–42. https://doi.org/10.1016/j.resconrec.2017.01.018
54. Kumar A, Samadder SR (2017) A review on technological options of waste to energy for effective management of municipal solid waste. Waste Manage 69:407–422. https://doi.org/10.1016/j.wasman.2017.08.046
55. Kumari A, Jha MK, Singh RP (2016) Recovery of metals from pyrolysed PCBs by hydrometallurgical techniques. Hydrometallurgy 165:97–105. https://doi.org/10.1016/j.hydromet.2015.10.020
56. Lacerda VG et al (2009) Separation of Cd and Ni from Ni-Cd batteries by an environmentally safe methodology employing aqueous two-phase systems. J Power Sources 193(2):908–913. https://doi.org/10.1016/j.jpowsour.2009.05.004
57. Laubertová M et al (2019) Metals recovery: Study of the kinetic aspects of copper acidic leaching waste printed circuit boards from discarded mobile phones. Acta Montanistica Slovaca 24(3):223–233
58. Laubertová M\, Novický M, Vindt T (2011) The possibilities of hydrometallurgical treatment of discarded mobile phones, pp 412–418
59. Lee H, Mishra B (2018) Selective recovery and separation of copper and iron from fine materials of electronic waste processing. Miner Eng 123(December 2017):1–7. https://doi.org/10.1016/j.mineng.2018.04.021
60. Lee JC, Song HT, Yoo JM (2007) Present status of the recycling of waste electrical and electronic equipment in Korea. Resour Conserv Recycl 50(4):380–397. https://doi.org/10.1016/j.resconrec.2007.01.010

61. Lekka M et al (2015) Gold recovery from waste electrical and electronic equipment by electrodeposition: a feasibility study. Hydrometallurgy 157:97–106. https://doi.org/10.1016/j.hydromet.2015.07.017
62. Li H, Eksteen J, Oraby E (2018) Hydrometallurgical recovery of metals from waste printed circuit boards (WPCBs): current status and perspectives—a review. Resour Conserv Recycl 139(August):122–139. https://doi.org/10.1016/j.resconrec.2018.08.007
63. Li J et al (2007) Recycle technology for recovering resources and products from waste printed circuit boards. Environ Sci Technol 41(6):1995–2000. https://doi.org/10.1021/es0618245
64. Li Z et al (2019) Comparative life cycle analysis for value recovery of precious metals and rare earth elements from electronic waste. Resour Conserv Recycl 149(March):20–30. https://doi.org/10.1016/j.resconrec.2019.05.025
65. Lisińska M, Saternus M, Willner J (2018) Research of leaching of the printed circuit boards coming from waste mobile phones. Arch Metall Mater 63(1):143–147. https://doi.org/10.24425/118921
66. Liu Y et al (2018) Impact of the operating conditions on the derived products and the reaction mechanism in vacuum pyrolysis treatment of the organic material in waste integrated circuits. J Clean Prod 197:1488–1497. https://doi.org/10.1016/j.jclepro.2018.05.236
67. Long L et al (2010) Using vacuum pyrolysis and mechanical processing for recycling waste printed circuit boards. J Hazard Mater 177(1–3):626–632. https://doi.org/10.1016/j.jhazmat.2009.12.078
68. Ma L et al (2013) Theoretical simulation and experimental study on nickel, cobalt, manganese separation in complexation-precipitation system. Separ Purification Technol 108:124–132. https://doi.org/10.1016/j.seppur.2013.01.044
69. Ma Y et al (2021) A promising selective recovery process of valuable metals from spent lithium ion batteries via reduction roasting and ammonia leaching. J Hazard Mater 402(July 2020):123491. https://doi.org/10.1016/j.jhazmat.2020.123491
70. March JA et al (2004) Fibromatosis: Abordaje de York-Mason modificado para la escisión de un tumor desmoide en la región sacra. Cirugia Espanola 76(3):187–190. https://doi.org/10.1016/S0009-739X(04)78961-7
71. Moher D et al (2009) Preferred reporting items for systematic reviews and meta-analyses: the PRISMA statement. PLoS Med 6(7). https://doi.org/10.1371/journal.pmed.1000097
72. Moltó J et al (2011) Thermal decomposition of electronic wastes: mobile phone case and other parts. Waste Manage 31(12):2546–2552. https://doi.org/10.1016/j.wasman.2011.07.028
73. Nekouei RK et al (2018) Two-step pre-processing enrichment of waste printed circuit boards: mechanical milling and physical separation. J Clean Prod 184:1113–1124. https://doi.org/10.1016/j.jclepro.2018.02.250
74. Ning C et al (2017) Waste printed circuit board (PCB) recycling techniques. Top Curr Chem 375(2):21–22. https://doi.org/10.1007/s41061-017-0118-7
75. Oishi T et al (2007) Recovery of high purity copper cathode from printed circuit boards using ammoniacal sulfate or chloride solutions. Hydrometallurgy 89(1–2):82–88. https://doi.org/10.1016/j.hydromet.2007.05.010
76. Ongondo FO, Williams ID, Cherrett TJ (2011) How are WEEE doing? A global review of the management of electrical and electronic wastes. Waste Manage, 714–730. https://doi.org/10.1016/j.wasman.2010.10.023
77. Park YJ, Fray DJ (2009) Recovery of high purity precious metals from printed circuit boards. J Hazard Mater 164(2–3):1152–1158. https://doi.org/10.1016/j.jhazmat.2008.09.043
78. Petter PMH, Veit HM, Bernardes AM (2014) Evaluation of gold and silver leaching from printed circuit board of cellphones. Waste Manage 34(2):475–482. https://doi.org/10.1016/j.wasman.2013.10.032
79. Pietrelli L, Ferro S, Vocciante M (2019) Eco-friendly and cost-effective strategies for metals recovery from printed circuit boards. Renew Sustain Energy Rev 112(June):317–323. https://doi.org/10.1016/j.rser.2019.05.055
80. Polák M, Drápalová L (2012) Estimation of end of life mobile phones generation: the case study of the Czech Republic. Waste Manage [Preprint]. https://doi.org/10.1016/j.wasman.2012.03.028

81. Priya A, Hait S (2017)'Comparative assessment of metallurgical recovery of metals from electronic waste with special emphasis on bioleaching. Environ Sci Pollution Res 24(8):6989–7008. https://doi.org/10.1007/s11356-016-8313-6
82. Qu Y et al (2019) Understanding residents' preferences for e-waste collection in China—a case study of waste mobile phones. J Clean Prod 228:52–62. https://doi.org/10.1016/j.jclepro.2019.04.216
83. Rao MD et al (2020) Challenges and opportunities in the recovery of gold from electronic waste. RSC Adv 10(8):4300–4309. https://doi.org/10.1039/c9ra07607g
84. Rao MD et al (2021) Optimization of process parameters for the selective leaching of copper, nickel and isolation of gold from obsolete mobile phone PCBs. Clean Eng Technol 4:100180. https://doi.org/10.1016/j.clet.2021.100180
85. Rene ER et al (2021) Electronic waste generation, recycling and resource recovery: technological perspectives and trends. J Hazard Mater 416(May 2020):125664. https://doi.org/10.1016/j.jhazmat.2021.125664
86. Ribeiro PPM, Dos Santos ID, Dutra AJB (2019) Copper and metals concentration from printed circuit boards using a zig-zag classifier. J Mater Res Technol 8(1):513–520. https://doi.org/10.1016/j.jmrt.2018.05.003
87. Riedewald F, Sousa-Gallagher M (2015) Novel waste printed circuit board recycling process with molten salt. MethodsX 2:100–106. https://doi.org/10.1016/j.mex.2015.02.010
88. Rocchetti L et al (2013) Environmental impact assessment of hydrometallurgical processes for metal recovery from WEEE residues using a portable prototype plant. Environ Sci Technol 47(3):1581–1588. https://doi.org/10.1021/es302192t
89. Rubin RS et al (2014) Utilization of life cycle assessment methodology to compare two strategies for recovery of copper from printed circuit board scrap. J Clean Prod 64:297–305. https://doi.org/10.1016/j.jclepro.2013.07.051
90. Sadegh Safarzadeh M et al (2007) A review on hydrometallurgical extraction and recovery of cadmium from various resources. Miner Eng 20(3):211–220. https://doi.org/10.1016/j.mineng.2006.07.001
91. Sarvar M, Salarirad MM, Shabani MA (2015) Characterization and mechanical separation of metals from computer printed circuit boards (PCBs) based on mineral processing methods. Waste Manage 45:246–257. https://doi.org/10.1016/j.wasman.2015.06.020
92. Sayar NA, Filiz M, Sayar AA (2009) Extraction of Co(II) and Ni(II) from concentrated HCl solutions using Alamine 336. Hydrometallurgy 96(1–2):148–153. https://doi.org/10.1016/j.hydromet.2008.09.005
93. Sayilgan E et al (2009) A review of technologies for the recovery of metals from spent alkaline and zinc-carbon batteries. Hydrometallurgy 97(3–4):158–166. https://doi.org/10.1016/j.hydromet.2009.02.008
94. Senanayake G (2005) Gold leaching by thiosulphate solutions: a critical review on copper(II)-thiosulphate-oxygen interactions. Miner Eng 18(10):995–1009. https://doi.org/10.1016/j.mineng.2005.01.006
95. Sepúlveda A et al (2010) A review of the environmental fate and effects of hazardous substances released from electrical and electronic equipments during recycling: examples from China and India. Environ Impact Assess Review [Preprint]. https://doi.org/10.1016/j.eiar.2009.04.001
96. Sethurajan M et al (2019) Recent advances on hydrometallurgical recovery of critical and precious elements from end of life electronic wastes—a review. Crit Rev Environ Sci Technol 49(3):212–275. https://doi.org/10.1080/10643389.2018.1540760
97. Sheng PP, Etsell TH (2007) Recovery of gold from computer circuit board scrap using aqua regia. Waste Manage Res 25(4):380–383. https://doi.org/10.1177/0734242X07076946
98. Silva WC et al (2018) Recovery of base metals, silicon and fluoride ions from mobile phone printed circuit boards after leaching with hydrogen fluoride and hydrogen peroxide mixtures. Waste Manage 78:781–788. https://doi.org/10.1016/j.wasman.2018.06.049
99. Silvas FPC et al (2015) Printed circuit board recycling: physical processing and copper extraction by selective leaching. Waste Manage 46:503–510. https://doi.org/10.1016/j.wasman.2015.08.030

100. Sinha-Khetriwal D, Kraeuchi P, Schwaninger M (2005) A comparison of electronic waste recycling in Switzerland and in India. Environ Impact Assess Rev 25(5 SPEC. ISS.). https://doi.org/10.1016/j.eiar.2005.04.006
101. Sommerville R et al (2021) A qualitative assessment of lithium ion battery recycling processes. Resour Conserv Recycl 165(October 2020):105219. https://doi.org/10.1016/j.resconrec.2020.105219
102. Statista (2021) Forecast number of mobile users worldwide 2020–2025
103. Tan Q et al (2018) Rethinking residential consumers' behavior in discarding obsolete mobile phones in China. J Clean Prod 195:1228–1236. https://doi.org/10.1016/j.jclepro.2018.05.244
104. Torihara K, Kitajima T, Mishima N (2015) Design of a proper recycling process for small-sized e-waste. Procedia CIRP 26:746–751. https://doi.org/10.1016/j.procir.2014.07.088
105. Torres R, Lapidus GT (2017) Closed circuit recovery of copper, lead and iron from electronic waste with citrate solutions. Waste Manage 60:561–568. https://doi.org/10.1016/j.wasman.2016.12.001
106. Tripathi A et al (2012) Leaching of gold from the waste mobile phone printed circuit boards (PCBs) with ammonium thiosulphate. Int J Metall Eng 1(2):17–21. https://doi.org/10.5923/j.ijmee.20120102.02
107. Tuncuk A et al (2012) Aqueous metal recovery techniques from e-scrap: hydrometallurgy in recycling. Miner Eng 25(1):28–37. https://doi.org/10.1016/j.mineng.2011.09.019
108. Vats MC, Singh SK (2015) Assessment of gold and silver in assorted mobile phone printed circuit boards (PCBs): original article. Waste Manage 45:280–288. https://doi.org/10.1016/j.wasman.2015.06.002
109. Ventura E et al (2018) Physical and thermal processing of waste printed circuit boards aiming for the recovery of gold and copper. J Environ Manage 223(February):297–305. https://doi.org/10.1016/j.jenvman.2018.06.019
110. Wang F et al (2015) Mineralogical analysis of dust collected from typical recycling line of waste printed circuit boards. Waste Manage 43:434–441. https://doi.org/10.1016/j.wasman.2015.06.021
111. Wang H et al (2017a) Recovery of waste printed circuit boards through pyrometallurgical processing: a review. Resour Conserv Recycl 126(August):209–218. https://doi.org/10.1016/j.resconrec.2017.08.001
112. Wang H et al (2017b) Recovery of waste printed circuit boards through pyrometallurgical processing: a review. Resour Conserv Recycl 126(April):209–218. https://doi.org/10.1016/j.resconrec.2017.08.001
113. Wansi E et al (2018) Waste management of discarded cell phones and proposal of material recovery techniques. Procedia CIRP 69(May):974–979. https://doi.org/10.1016/j.procir.2017.11.011
114. Williams PT (2010) Valorization of printed circuit boards from waste electrical and electronic equipment by pyrolysis. Waste and Biomass Valorization 1(1):107–120. https://doi.org/10.1007/s12649-009-9003-0
115. Willner J, Fornalczyk A, Saternus M (2015) Selective recovery of copper from solutions after bioleaching electronic waste. Nova Biotechnologica et Chimica 14(1):32–37. https://doi.org/10.1515/nbec-2015-0011
116. Wu Q et al (2019) Trace metals in e-waste lead to serious health risk through consumption of rice growing near an abandoned e-waste recycling site: comparisons with PBDEs and AHFRs. Environ Pollution 247:46–54. https://doi.org/10.1016/j.envpol.2018.12.051
117. Wu Z et al (2017) A critical review on the recycling of copper and precious metals from waste printed circuit boards using hydrometallurgy. Front Environ Sci Eng 11(5). https://doi.org/10.1007/s11783-017-0995-6
118. Xiu FR et al (2017) A novel recovery method of copper from waste printed circuit boards by supercritical methanol process: preparation of ultrafine copper materials. Waste Manage 60:643–651. https://doi.org/10.1016/j.wasman.2016.11.001
119. Yang H, Liu J, Yang J (2011) Leaching copper from shredded particles of waste printed circuit boards. J Hazard Mater 187(1–3):393–400. https://doi.org/10.1016/j.jhazmat.2011.01.051

120. Yin J, Gao Y, Xu H (2014) Survey and analysis of consumers' behaviour of waste mobile phone recycling in China. J Clean Prod 65:517–525. https://doi.org/10.1016/j.jclepro.2013.10.006
121. Zhang XM, Senanayake G (2016) A review of ammoniacal thiosulfate leaching of gold: an update useful for further research in non-cyanide gold lixiviants. Miner Process Extractive Metall Rev 37(6):385–411. https://doi.org/10.1080/08827508.2016.1218872
122. Zhang Y et al (2012) Current status on leaching precious metals from waste printed circuit boards. Procedia Environ Sci 16:560–568. https://doi.org/10.1016/j.proenv.2012.10.077
123. Zheng G et al (2013) Association between lung function in school children and exposure to three transition metals from an e-waste recycling area. J Exposure Sci Environ Epidemiol 23(1):67–72. https://doi.org/10.1038/jes.2012.84

Challenges and Opportunities of a Circular Economy Approach to Sustainable Recycling of Spent Lithium-Ion Batteries (LIBs): A Review

O. Saba-Bawazir, Meltem Alkoyak, R. Bhavani Rao, Santosh Mahapatra, and Keshavan Varadarajan

Abstract Lithium-ion batteries are widely implemented in applications and areas such as personal portable devices (PPD) and electric vehicles (EVs). Based on the rapid manufacturing development of LIBs, besides the notable increase in PPD consumption during COVID-19, limited regions, the rising prices of raw materials of LIBs, and the irresponsible disposal of LIBs in landfills cause a serious high risk to the environment and human health. Because of this, recovering precious materials from spent LIBs can solve the problem of limited available resources and mitigate environmental and human health impacts. This review aims to critically assess the available information on different aspects of recycling spent LIBs. In addition, a brief review of policies and regulations for some developed and developing countries will be provided, and it then examines the waste mobile phones, the customers' perceptions of disposal and recycling spent LIBs, the human health and environmental impacts of spent Lithium-ion batteries (LIBs) upon disposal and recycling, the economic aspects

O. Saba-Bawazir (✉)
Amrita School for Sustainable Development, Faculty of Interdisciplinary Studies, Amrita Vishwa Vidyapeetham University, Amritapuri, India
e-mail: Sababawazir@yahoo.com

Chemical Safety and Waste Management Department, Environmental Protection Authority, Ministry of Water and Environment, Aden, Republic of Yemen

M. Alkoyak
Department of Cognitive Sciences and Psychology, Amrita Vishwa Vidyapeetham University, Amritapuri, India

R. B. Rao
Center for Women's Empowerment and Gender Equality, School for Social and Behavioural Sciences, Amrita Vishwa Vidyapeetham University, Amritapuri, India

S. Mahapatra
Operations Management and Information Systems, School of Business, Clarkson University, Potsdam, NY, USA

K. Varadarajan
Samsung Electronics, Austin, USA

A. Cavalcante da Silva Batalhão and P. Singh (eds.), *Blending Approaches Towards Sustainability Focus*, Earth and Environmental Sciences Library,
https://doi.org/10.1007/978-3-031-93518-3_4

of recycling spent LIBs, the circular economy for spent LIBs, and the conventional and sustainable recycling processes on a commercial and laboratory scales will be discussed. Due to the lack of published review papers that covered all aspects of sustainable recycling processes spent LIBs, economic feasibility, circular economy, and environmental impact analysis, this review concludes with recommendations on future research the need for a holistic approach to achieving the circular economy strategies of spent LIBs recycling processes. Furthermore, this study will be helpful to scholars, policy makers, and producers to achieve the long-term sustainability of lithium-ion batteries.

Keywords Spent lithium-ion batteries (LIBs) · Recycling processes · Portable consumer electronic devices (PCEDs) · Economic benefits · Circular economy · Environmental and human health impacts

1 Introduction

Whereas the Non-Organisation for Economic Co-operation and Development (OECD) countries such as China and India are expected to cause global energy demand to increase by 90% due to population growth, rapid urbanization, advances in economic activities, and technological advances [98]. Greenhouse gas emissions and pollutants resulting from the extensive use of fossil fuels are primarily responsible for environmental issues and climate changes [66, 107]. In the past 20 years, the search for alternative energy sources has been extensive. Nevertheless, a majority of renewable energy sources exhibit natural intermittency, and the consequent storage systems from those sources are essential for continuous energy supply. Therefore, batteries are one of the frequently utilized systems for energy storage [143]. Currently, there has been a massive increase in the demand for Lithium-ion batteries owing to a host of good attributes over other kinds of batteries [69, 102]. Some of these positives include; low self-discharge, high energy density, and high faradaic efficiency [94, 153], and they are widely implemented in many different applications and areas [131], such as personal portable devices (laptops, tablets, mobile phones, digital cameras) [74, 101, 136, 149], medical technology [62], military [26], electric vehicles (EVs) [70, 114]. Nevertheless, usage of LIBs is expected to increase by leaps and bounds shortly due to their use in EVs [57, 142]. The use of LIBs is expected to grow at a compounded annual growth rate of 18% from 2020 to 2027 [2]. Therefore, several governments of developed and developing countries such as the United Kingdom, France, the United States, Germany, Italy, Spain, Norway, Sweden, India, China, Japan, and South Korea have released rules to promote electric vehicles sales [27, 88, 154], which lead to 80% estimated increase in the market share of LIBs by 2030. In addition, as reported by Bloomberg New Energy Finance (BNEF), the global consumption of raw materials of LIBs such as lithium, cobalt, and copper are projected to exceed by 20 times in 2030. Furthermore, supplies and reserves of the natural reserves of those raw materials are mostly in limited regions of the world such as Congo, Australia, Brazil, Russia, Argentina, Chile, Bolivia, Chile, Peru, and China [44]. On the other hand, for example, 25% of Co is used to produce LIBs for global demand, which causes the loss of precious materials such as Co [113]. Limited available resources in the natural reserves of lithium make the cost of the lithium unit for LIBs gradually increase globally [29, 47]. Based on the rapid manufacturing development of LIBs, and besides the increased consumption, limited regions and the rise prices of raw materials of LIBs, have thus resulted directly threatens and challenges to their sustainability. However, spent LIBs contain valuable rare earth materials such as lithium (Li), cobalt (Co), copper (Cu), nickel (Ni) [71, 134–141], which are limitedly extracted from spent LIBs in few countries [85]. Due to spent LIBs containing toxic chemical composition, such as reactive materials and organic materials, make them a problem [42], mainly after they reach their end of life (EOL), and LIBs have a particular lifetime. They may be leaking use, bulging, or degrading after repeated use [129]. Irresponsible disposal of LIBs in landfills causes a high risk to the environment and human health by explosions seriously and contaminating soil and groundwater [19, 39, 93]. Therefore, recovering precious materials from spent LIBs can not only solve the problem of limited available resources but also mitigate environmental and human health impacts [115]. Thus, it is broadly acknowledged that recycling spent LIBs is crucial for the long term growth of the LIBs industry. In the past five years, recent research published papers significantly devoted to recovery and recycling precious materials from spent LIBs focusing on developing technologies with limited attention on environmental and economic impacts assessment [143]. Furthermore, keeping people indoors and on devices during COVID-19 increases PC and other IT accessory sales [109]. High numbers of LIBs strongly penetrate the waste stream due to the widespread adoption of portable electronic devices [118]. Briefly, this is reflected in limited numbers of published papers. About 87 research papers and 7 review papers from Google Scholar, ScienceDirect, and Web of Science search engines on "Lithium-ion battery, Recycling processes, Circular Economy, Economic Benefits." Additionally, a holistic policy in manufacturing and recycling LIBs is crucial to ensuring a sustainable industry of LIBs. While there are many benefits in the technological advancement of recycling processes to extract valuable materials from spent LIBs, these developments will significantly restrict their practical implementation if the environmental impacts and economic feasibility are not considered [106, 118]. Despite there are enormous published research papers on various aspects of spent LIBs recycling processes, it appeared that there is a lack of published review papers which covered all aspects of spent LIBs sustainable recycling processes, economic feasibility, circular economy, and environmental impact analysis. This review study focuses on spent lithium-ion batteries from portable devices such as mobile phones and aims to critically evaluate the information that is currently available on those aspects. This review starts with a brief of policies and regulations for some developed and developing countries such as (USA, EU, Switzerland, China, Japan, South Korea, India, Australia). It then examines the waste mobile phones production and generation. Sections 4, 5, 6 and 7 presents analyse on customers perceptions of disposal and recycling spent LIBs, and the environmental and human health impacts of spent LIBs upon disposal and recycling, and economic aspects of recycling spent LIBs and circular economy for spent LIBs. Section 8 examines the conventional and sustainable recycling processes on a commercial and laboratory scales, while emphasizing the importance for a comprehensive approach to the sustainable recycling processes of spent LIBs to effectively achieve the circular economy strategies.

2 Current Policies and Regulations of Spent Lithium-Ion Battery Management

In both consumer electronic devices and electronic vehicles (EVs), LIBs are growing rapidly, increasing the number of spent LIBs in the waste stream. While LIBs contain precious and hazardous materials, which create significant management issues and adverse impacts on the environment and human health after they reach their end of life if they are not adequately managed [143]. Thus, electronic waste management regulations are imperative for minimizing environmental impacts and protecting human health [139]. This section reviews some current regulations of lithium-ion battery management in the United States of America (USA), the European Union (EU), Switzerland, China, Japan, South Korea, India, and Australia.

2.1 USA

In the U.S., the Resource Conservation and Recovery Act (RCRA) regulated solid and hazardous wastes. The RCRA classified hazardous waste in two terms as listed or characteristic. The listed hazardous waste is therefore directly defined under RCRA, whereas, after a test, hazardous waste can be characterized to figure out whether it is hazardous in nature or not. Despite this, there is no current federal policy or regulations established by the Environmental Protection Agency (EPA) to list LIBs with other types of rechargeable batteries (such as nickel-cadmium and lead-acid) as hazardous wastes and to encourage recycling of spent LIBs [30, 110], co-regulatory regulations are implemented in certain states within the United States to manage the human health and environmental impacts associated with the inappropriate disposal of spent LIBs [15] and spent LIBs are treated and categorized under universal waste standards [30, 111, 133, 134]. Over 97% of the waste from lead-acid batteries produced in the United States (US) has been undergone recycling since the early 1990s [132] because the US government legislation has partially promoted the recycling of Lead-acid and Nickel-Cadmium batteries in order to mitigate the environmental and human health risks associated with the improper disposal of these battery types [38]. On the other side, few states have banned the disposal of spent LIBs from landfills. Instead, based on the type of the batteries, Environmental Handling Fees (EHF) have been mandated to batteries and higher charges for those

more if compared to the spent LIBs coming from EVs, industrial sectors are targeted at recycling spent portable LIBs [147]. Moreover, Mossali et al. [92] have ensured that in the last ten years, much more growth in rechargeable LIBs applications has currently been seen, where rechargeable LIBs are considered the prominent energy storage technology in portable electronic devices.

4 Customers Perceptions of Disposal and Recycling Spent LIBs

While China is currently regarded as one of the world's largest producers and consumers of lithium-ion batteries (LIBs) [43], most LIBs are utilized in consumer electronic (CE) or portable electronic devices [134]. Moreover, the increasing production of Electric Vehicles (EVs) has led to a rapid growth consumption and waste stream of LIBs [41], where LIBs sales are predicted to increase from 5 million in 2015 to 7 million in 2020. Thus, it is anticipated that this figure will reach 180 million by the year 2045 [40]. Despite the fact that the production of LIBs is mainly concentrated in Asian countries, where Japan accounts for 40% of the global production, followed by South Korea and China, and the major consumption rates are recorded at 28.4% and 27.2% in the USA and EU, respectively [41].

Furthermore, since this type of electronic waste is considered low-value e-waste or trashes, LIBs are not intentionally collected in China [43]. As a result, only 30–50% of the population is properly disposed of their spent LIBs of portable electronic devices globally, with unaware of the potential harmfulness of spent LIBs [120], versus 15.9% that disposed of their LIBs in the trash bins, and 59.6% that who store them at home [134].

A survey conducted by [43] in the top ten urban centers ranked by annual gross domestic product (GDP) in China shows that most participants, specifically 93.8%, were not aware of the proper disposal methods of their spent lithium-ion batteries. The study also found that spent LIBs have not been included in the current waste electronic and electrical equipment (WEEE) collection system. As a result, only a few committed collectors were found to recycle spent LIBs, which is currently estimated that less than 10% or even less than 5% of used spent LIBs in consumer electronics are collected and recycled. At the same time, most of them are always ended up in the landfill. On the other hand, Zhou and Xu [155] ensured that the recycling practices of spent LIBs remain not well-known despite some recycling companies being identified within China. Sun et al. [121] indicated that, due to the high rate associated with spent LIBs, a significant majority of the participants are concerned with the environmental issues, where Gu et al. [43] stated that 59.6% of the participants stored their spent LIBs at home. With little available information about spent LIBs from consumer electronics in China, understand the current situation of recycling LIBs from consumer electronics in other developing countries such as India. Because as reported by [36], almost 700 million users will start to use mobile internet in Asian Pacific, and around two-thirds of them will be from China and India by 2025. As a result, in terms of active unique smartphone users (1041 million users), India would be the second largest smartphone market in Asian Pacific, after China by 2025 [36]. In order to evaluate the various characteristics, attributes, and profiles of Chinese citizens' decisions between various waste mobile phone collection services, Qu et al. [104] conducted a conjoint analysis as a scenario tool. They found that the collecting price and quality certification were the most significant attributes. On the other hand, Nixon et al. [95] used the same method to analyse the different recycling programs of Californian residents' preferences. The results show that the collection cost and distance are the most significant factors. As well, many studies emphasized that the economic benefits (collection price) and convenience (collection distance) have played a main roles in a collection services for e-waste disposal in many countries [100, 104, 135, 146].

In contrast, 63.7% of mobile phone consumers would rather not prefer to take part in a waste collection program for their discarded mobile phones, as reported by [124] due to data security and recycling quality considerations. According to the findings of [7], data security is significant in participating in mobile phone collection waste programs. Mishima and Nishimura [90] stressed that waste mobile phone collection services have to ensure and promote data safety insurance to encourage waste collection participants in mobile phones. In order to attract mobile phone users, waste mobile phone collection services face a challenge in meeting customer needs and selecting the best trade-off of various preferences of their services. Therefore, the collection of electronic waste must be promoted as much as possible to support the conservation of natural resources and reduce the amount of solid waste. In particular, formal collection sectors need to pay more attention to quality certification, cost collection, distance collection, and data security insurance in fast-growing e-waste production and generation countries such as China and India.

5 Effects of Spent LIBs on the Environmental and Human Health upon Disposal and Recycling

While LIBs are considered a clean and renewable energy source, in reality, there are obvious potential environmental and human health impacts at the end of their life cycles, especially during disposal and recycling [143]. Moreover, the massive global growth of LIBs in battery markets has paid attention to understanding the environmental impacts associated with the disposal or recycling of LIBs at the end of their life cycles (Winslow et al., 2018a). Some studies have reported that, despite high metals content presented in different types of LIBs, only a small proportion of these metals are projected to leach in a landfill from discarded LIBs [60, 72, 110, 138]. On the other hand, most leachable metals such as lithium, cobalt, and nickel have contaminated soil and groundwater both inside and outside the landfills [43, 53, 71], posing a threat to the human health and environment if the leachate metals

Table 1 Different disposal methods and their associated environmental and human health impacts

Disposal method	Environmental impacts	Human health impacts
Landfills	Contaminating soils and groundwater [43, 71]	• Lithium targets the central nervous [76] • Free cobalt may cause cancer • Copper, iron, and nickel are all involved in the generation of reactive oxygen species, causing premature aging and DNA damage [87, 138] • Thallium has high human toxicity and can be easily absorbed through the skin [60, 103]
Incineration	Emission of toxic gases such as hydrogen fluoride (HF) [48]	Emitted toxic gases cause fire explosions and expose humans to serious injuries [120, 138]

are not properly treated [28]. Where lithium is known to target the human central nervous system [76]. Additionally, free cobalt has the potential to cause cancer to humans. Copper, iron, and nickel are all involved in the generation of reactive oxygen species, which is known to cause DNA damage and premature aging [87, 138], and Thallium has high human toxicity than lead, mercury, or cadmium, and can be easily absorbed through the skin [60, 103]. Furthermore, in the case of overheating or incineration, LIBs emit toxic gases like hydrogen fluoride (HF) [48], cause fire explosions, and expose people to serious injuries due to the improper recycling of spent LIBs, particularly with the presence of Li and a flammable electrolyte in high contact with moisture [120, 138].

Along with the rapid production of new LIBs generation [9], LIBs are facing production challenges, as well as the amount of spent LIBs is increasing and becoming an urgent problem to save natural reserves, protect the humans and the environment from harmful impacts [92]. Therefore, during the disposal or recycling of spent LIBs, care must significantly be taken to reduce the potential release of hazardous materials to the ecosystem. Table 1 demonstrates two different disposal methods and their associated environmental and human health impacts.

6 Economic Aspects of Recycling Spent LIBs

Although numerous models are being created to anticipate the economic and environmental impacts of various LIBs phases from the cradle to the grave [143], LIBs recycling processes are still at an early stage. Limited numbers are used commercially, and rare economic data are available [133]. Limited reviews addressed the economic issues of LIBs recycling techniques [143]. Moreover, LIBs recycling processes involve various operations starting with the collection from wide areas before being transported to the recovery facilities. LIBs recycling processes therefore as recognized are highly costed. On the other hand, recycling of spent LIBs is still seen as an incentive to mitigate the environmental impacts of the adverse disposal of spent LIBs and to reduce the economic impacts by ensuring the continuous supply of precious and rare materials from spent LIBs and as well alleviating price fluctuations [86]. Therefore, to establish a recycling business, recycling operations must be economically viable commercially [143]. Wang et al. [133, 134] conducted their studies to analyze the recycling cost of LIBs sensitively, and their studies found that recovery technology and the chemical composition of the precious materials in spent LIBs were the factors in the recycling process' profitability. For example, [38, 111, 143] are notably discovered that recycling of $LiCoO_2$ and LCO batteries has the highest economic value and revenue compared to other spent LIBs, which not or contained less precious materials such as Lithium, Cobalt or Nickel. Moreover, Dai et al. [23] has intensively emphasized that the required cost and the revenue generated are the major two factors in determining the profitability of the recycling processes, where the revenue generated of spent LIBs batteries are similar due to the regulated sales of the global trading markets of the metals. The required costs of recycling processes are significantly varied from one country to another. For example, Yang [143] stated that, compared to China and South Korea, the labor's cost and the capital costs of recovery plants for recycling spent LIBs are mainly more expensive in the US. Further, Olsson et al. [99] studied the revenue generated by recycling 10,000 tonnes of spent LIBs of the same chemical composition by different recycling techniques. The Pyrometallurgical, Hydrometallurgical, and Direct recycling processes were used to compare the revenue generated. The study found that the direct regeneration process produced the highest revenue and was considered the most economical recovery technique compared to the other two recovery techniques. Despite, the Hydrometallurgical process generated more revenue and large pure materials would be recovered compared to the pyrometallurgical process, but due to its many processing steps compared to pyrometallurgical process, hydrometallurgical process often does not prefer. Moreover, the recycling profitability of spent LIBs is highly relied on the transportation cost as well [50], which is five and ten times higher in the US than in China and South Korea [99]. Therefore, due to the few global processing facilities found on a commercial scale to recycle spent LIBs, Tabassum et al. [123] significantly recommend that much more developments should be done in the commercial viability of recycling spent LIBs. In addition, for future research, both [3, 118] have currently emphasized the importance of studying the economic analysis of recycled materials from the electronic wastes such the mobile phones by developed sustainable recycling processes, to improve management policies and legalisation of electronic waste and spent LIBs to help countries achieve the green environment and save human health (Fig. 1 and Table 2).

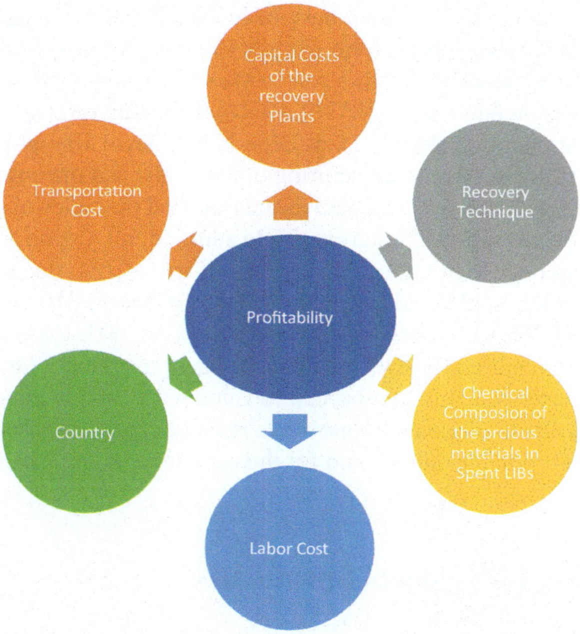

Fig. 1 Main factors determine the profitability of established recycling spent LIBs business [23, 133, 134, 143]

Table 2 Compression of recycling 10,000 tones sample of spent LIBs by pyrometallurgical, hydrometallurgical, and direct recycling processes [99]

10,000 tones sample of spent LIBs	Direct regeneration process	Hydrometallurgical process	Pyrometallurgical process
Generated revenue	High	Medium	Low
Economic recovery techniques	High	Medium	Low
Pure recovery materials	High	Medium	Low
Processing steps	Less processing steps	Long processing steps	Medium processing steps

7 Circular Economy for Spent LIBs

The Circular Economy (CE) is designed as a regenerative strategy to decrease waste and ensure the environmental sustainable disposal of discarded products [148]. As it is Acknowledged, China and South Africa are the major Li natural reserves in the world, where roasting and leaching processes are used to extract Li from igneous rocks [89]. In fact, due to technological obstacles, the extraction of Li from brine is limited, where the extraction of 1 ton of Li requires 20,000 tons of water [61]. Li's demand is therefore projected to be higher than the mining supply in 2023–2025 [119]. In addition, the 25% of cobalt (Co) used for the total request of the manufacturing of LIBs is extracted by deposits from the Democratic Republic of the

Congo (DRC), which have been sanctioned due to human rights abuses, and while DRC owns 46% of the world's reserves of Co and contributes 69% of the EU with reserves of Co [84]. Even worse, 170 folds of the current extracting capacity are required to extract Ni [96]. Due to the nature reserves are limited and 90% efficiency are not recycled of spent LIBs [119], new LIBs with less cobalt are produced [9]. However, the manufacturing of new LIBs and the unstable legislative framework are significantly represented as the practical obstacles of innovative recycling processes to industrial exploitation [9]. In addition, relying on achieving the circular economic strategy to promote the recycling of spent LIBs, Baars et al. [5] stressed that 30–90% of the total cobalt demand for the production LIBs will be supplied from recycling by 2050. Furthermore, 260 k-tons of Co, 250 k-tons of Li, and 1300 k-tons of Ni will be required for EVs batteries production annually [46]. Therefore, [9, 92, 118] have ensured that it is important to value the available materials in the next generation of LIBs to implement innovative recycling processes in order to implement effectively a profitable value chain for the new generation of LIBs in the circular economy.

8 Recycling of Spent LIBs

Although most literature significantly reviews from different perspectives the recycling of spent LIBs, all agree that much more development is needed to achieve sustainable and economic benefits for all types of LIBs. Whereas the majority of commercial recycling companies use one or both processes of mechanical recycling, such as hydrometallurgy and pyrometallurgy, focusing on extracting the high precious materials, rather than the implementation of closed-loop recycling for other component materials. This approach neglects the technological and environmental gaps associated with the discarded component materials of spent LIBs and fails to the economic thinking of other component materials, as well as the comparison of various recycling techniques in the existing academic literature [118].

Therefore, as it is acknowledged, recycling processes of spent LIBs allow to mitigate environmental impacts, save natural reserves from mining raw materials and avoiding imports, minimize energy consumption and emissions of CO_2, discover economic gains, reduce waste and safely manage toxic materials from spent LIBs to extract precious materials with high recovery efficiency [8, 92].

Currently, despite the major motivation for the recycling spent LIBs is the precious materials contained in the cathodic active layer and their economic value, where their economic value represents 90% of the total value of spent LIBs [65] as reported in Table 3, nowadays less than 3% of spent LIBs are globally recycled [92, 120], and on the London Metal Exchange Market, the cathode materials values have rapidly changed, reached $95,000 per ton of cobalt (Co) in March 2018 and declined to $29,000 per ton in March 2019 [20]. Although lithium is currently not considered one of the critical materials list [22, 122], it is expected to increase in the production of many latest automotive chemistry batteries (Electronic Vehicles (EVs)) applications such as $Li[Ni, Co, Al]O_2$ (NCA) and $Li[Ni, Mn, Co]O_2$ (NMC) [20]. At the

Table 3 The prices of LIBs precious materials (Trading [92, 128])

Precious materials	Prices (USD/T)	
	2019	2021
Cobalt	35,500	42,535
Lithium	10,000	13,911.95
Copper	5800	8279.30
Aluminium	1800	2454.75
Manganese	2000	5.04
Graphite	800	200–1800

commercial level, the required costs and revenue generated are the two main drivers of recycling spent LIBs, so only Ni, Co, steel, Cu, and Al are currently recovered from spent LIBs, while Li, Mn, and graphite are rarely recovered, and plastics are burned for energy recovery [24, 92]. Despite this, cobalt and graphite are the main materials in LIBs and are considered mainly sufficient and exposed to supply chain risk, as reported by the European Commission in 2017 [34].

Consequently, only spent LIBs with high content of Cobalt (Co) and other precious materials are highly preferred for recycling, where the revenue generated from recycling Lithium Cobalt Oxide (LCO) and Nickel Manganese Cobalt (NMC) cathode batteries is 8900$ per metric ton, on the contrast, LMO has a low value about 860$ per metric ton cathode batteries and Lithium Iron Phosphate (LFP) has no valuable materials [133, 134, 138]. Wherein it was estimated that 13% of spent LIBs cost per kWh could be saved from metals recycling from spent LIBs [92, 120]. Furthermore, the recovered precious materials from recycling spent LIBs are used as raw materials in the production of other batteries to preserve raw materials [6]. Despite the predominant trend currently to reduce the LIBs production cost by replacing cobalt (Co) metal with other available metals, Chen [18] called to work on developing the recycling processes to recover metals from LIBs regardless of their limited content of metals to balance the necessary recycling costs with the effects generated revenue. Along with these, Diekmann [25] ensured that several recycling challenges should be considered to develop a flexible process such as the input chemistries variability, impurities presence, variability of geometric, the recent developments in the batteries market, the safe management of spent LIBs components due to exposure to lithium atoms, standardization, the scalability of processes, and the simplification of recovery steps. Therefore, Sommerville [118] concluded that the recycling of LIBs must be mainly crucial for future sustainability rather than an option due to several environmental and economic factors such as;

(1) Environmental concerns regarding the improper disposal and recycling of spent LIBs.
(2) Economic concerns regarding the vulnerable supply risk of certain critical raw materials lead to an increase in the cost of those materials for LIBs manufacturing.

8.1 Hydrometallurgical Processes

Li et al. [73, 75] ensured that the hydrometallurgy process depends on the solid to liquid ratio, duration of leaching, types of reductant, and temperatures. This technique facilitate the extraction of metals by high purity level and low energy requirements, and good efficiency. Moreover, Horeh et al. [52] stated that this process needs fewer operation requirements and low costs.

On the other hand, Boxall et al. [11] declared that the main issues of this process are the low concentration of sulfur (S) and Ferrum (Fe) in spent LIBs that leads to reaction activation, slow leaching times, and the pulp density required during the process. Furthermore, according to Mossali et al. [92], Zhao et al. [152], Yang et al. [144], this process is considered to be a more complex process that is strongly dependent on the cathode chemical contents, and more process steps are required, such as solvent extraction to separate the constituent metals, electrochemical processing, chemical precipitation, and base leaching or acids leading to the unsustainable industrial recovery of the precious metals from spent LIBs.

For example, Boyden et al. [12] conducted a study to compare the environmental impacts between hydrometallurgical and pyrometallurgical processes. The findings indicated that plastic incineration has caused the most considerable influence on human toxicity, global warming, and terrestrial ecotoxicity associated with the electricity generation in pyrometallurgical processes. On the other hand, the findings showed that the hydrometallurgical process has greater impacts than the pyrometallurgical process within the potential global warming. Gaines et al. [37] emphasized that while this method is considered more popular for recycling spent LIBs, it is important to have less extraneous materials such as plastics, electrolytes, casings, and graphite to ensure that this process is cost-effective [118].

8.2 Pyrometallurgy Processes

Although pyrometallurgy processes are classified as high capital cost processes on a commercial scale [56], pyrometallurgy is the most widely employed recycling technique for metals extraction from all types of spent LIBs through simple processes [56, 92, 130]. Moreover, high-temperature treatments lead significantly to the emission of harmful gases and require intensive energy consumption [92, 120]. For instance, at the industrial scale, Umicore's processing operations consumed energy of 5000 MJ to smelt a single ton of spent LIBs and to purify the emitted gases. In contrast, the Sony-Sumimoto's processing operations required 992 MJ per ton for the incineration of organic materials [120]. Although the fact that pyrometallurgy processes are 100% recyclable for all types of spent LIBs, due to the economic infeasibility of slagged leaching [105], pyrometallurgy is generally unable to recover Li [56, 92, 105]; however, researches at lithium extraction from slag continue to progress [117]. Moreover, the content of precious materials in spent LIBs such as cobalt is decreased, the profitability of the pyrometallurgical process is also reduced. In this context, substances including plastics, electrolyte, graphite, lithium, manganese, and aluminum are not extracted, but rather are found in the slag, which is subsequently used as an aggregate [37]. Mossali et al. [92] have called for pushing future research towards more effective and sustainable processes for the recycling of spent LIBs.

8.3 Bio-hydrometallurgical Processes

Bio-hydrometallurgical processes are considered at the laboratory scale as a minor alternative for recycling spent LIBs, in which electrode materials are dissolved and excreted respectively by metabolites and fungi or microorganisms. In addition, pure acids are used to improve dissolution [140]. Moreover, compared to other recycling processes, the microorganisms and metabolites used to extract the precious materials have much more advantages, such as performed at low temperatures, less emitted toxic gases, fewer energy inputs and operating costs required, and environmentally safe [59, 83]. However, despite its eco-friendliness (Economic and Environmental advantages) in the long run, reduces pollution, waste generated and low energy consumption [51, 73, 75, 140, 141], this process is still challenging and under investigation recently on a large-scale, mainly when it targets spent LIBs recycling [16, 92]. On the other hand, Yang et al. [143] ensured that this method is considerably more sustainable strategy to the recycling of spent LIBs; however, the slow kinetics associated with this process significantly influence operating costs and cycles when implementing in large-scale industrial applications. However, still, many studies reported that this process is sustainable, simple, safe, and has fewer industrial requirements comparing to the conventional processes such as hydrometallurgical and pyrometallurgy processes [6, 10, 116], and it will play a necessary roles in electronic waste metal recovery in future [54, 55] (Morin et al. 2006) (Table 4).

9 Conclusion and Recommendations for Future Researches

Currently, rechargeable lithium-ion batteries (LIBs) are widely implemented to store energy in many different applications and areas, such as personal portable devices, energy systems, and EVs. So many LIBs are estimated to increase by 80% in the market share of batteries by 2030. Along with the massive growth of electronic waste from personal portable electronic devices during COVID-19 globally, spent LIBs mainly have to be yet paid attention to understanding the environmental and economic impacts associated with the disposal or recycling of LIBs at the end of their life cycles. Furthermore, the global consumption of raw materials of LIBs such as lithium and cobalt is expected to increase by 2030. Supplies and reserves of the natural reserves of those raw materials are mostly in limited regions of the world. Therefore, there is an urgent need that these batteries can be effectively disposed of and recycled

Table 4 A comparison between the hydrometallurgy, pyrometallurgy, and bio-hydrometallurgy recycling techniques of lithium-ion batteries

Hydrometallurgy	Pyrometallurgy	Bio-hydrometallurgical
Industrial-scale (commercial level)	Industrial-scale (commercial level)	Laboratory scale (under investigation recently on a large-scale, mainly when it targets spent LIBs recycling)
Unsustainable	Unsustainable	Sustainable
Economic benefits	Economic benefits (economic infeasibility of slagged leaching)	Economic benefits
Focusing on extracting the high precious materials, low sulfur (S), and ferrum (Fe) in spent LIBs leads to reaction activation, slow leaching times, and the pulp density required during the process	Focusing on extracting the high precious materials (unable to recover Li plastics, electrolyte, graphite, Mn and Al are not extracted but are found in the slag which used as an aggregate)	Focusing on extracting the high precious materials and other component materials (**Closed-Loop**). The microorganisms and metabolites used to extract the precious materials have much more advantages, such as performed at low temperatures, less emitted toxic gases, less energy inputs, and being environmentally safe
Environmental impacts	Environmental Impacts	Eco-friendliness
Recovers metals with high purity, low energy requirements, and good efficiency	High-temperature treatments lead significantly to the emission of harmful gases and require intensive energy consumption	Electrode materials are dissolved and excreted by metabolites and fungi or microorganisms, and pure acids are used to improve dissolution, reduce pollution, waste generated, and low energy consumption
Less operation requirements and low costs	High capital cost processes on a commercial scale	Less operating costs required
A more complex process that is strongly dependent on the cathode chemical contents and more process steps are required such as solvent extraction to separate the constituent metals, electrochemical processing, chemical precipitation, and base leaching or acids leading to the unsustainable industrial recovery of the precious metals from spent LIBs	The most widely used recycling technique is to extract metals from all types of spent LIBs using simple processes	Slow kinetics has a significant impact on operating costs and cycles in large-scale industrial application

(continued)

Table 4 (continued)

Hydrometallurgy	Pyrometallurgy	Bio-hydrometallurgical
More popular for the recycling of spent LIBs	100% recyclable for all types of spent LIBs	Simple, safe, and has less industrial requirements comparing to conventional processes such as hydrometallurgical and pyrometallurgy processes

to facilitate the sustainable production of these batteries, and cocurrently reduce their waste and associated cost. This review comprehensively discussed the challenges and opportunities available in helping the sustainable management of spent lithium-ion batteries. Thus, under separate headings, these challenges and opportunities are summarised where these recommendations are strongly related to each other.

- **Policies and Legislations Development Aspects of Spent LIBs Management:**

1. In order to run and increase collection rates, and to continue encouraging the LIBs producers to establish their collection centers and recycling plant, and collaborate with other formal or informal collection centers, countries' governments' legislations have to promote sustainable recycling programs of spent LIBs in parallel with promoting EVs sales.
2. As LIBs are not yet classified as hazardous waste, more direct regulation is needed to connect the collection of spent LIBs to Extended Producer Responsibility (EPR), enhancing the economy and reducing environmental impacts to apply the effective sustainable recycling strategy for spent LIBs.
3. To effectively ban the importation of e-waste products and reduce the movement of toxic materials, it is crucial that collection and recycling rates are increased through the implementation of financial incentives in numerous developed and developing countries.
4. Strong policies have to be enforced or amended to develop new batteries to sustain the batteries industry, with precise classification, responsibilities, and guidelines for those developed lithium-ion batteries.
5. Most consumer electronic (CE) or portable electronic devices contain LIBs, which are increasing in the production of electric Vehicles (EVs), which have led to an increased waste stream of LIBs. In order to raise awareness of the potential harmfulness of spent LIBs, much more future research is required to study the awareness of these consumers' portable electronic devices and factors that influence the disposal of spent LIBs in different developing and developed countries.

- **Economic and Circular Economic Aspects of Recycling Spent LIBs:**

1. Limited numbers of spent LIBs recycling processes are used commercially, resulting in the current availability of rare economic data and limited reviews

covering the economic aspects of LIBs recycling processes. In this regard, LIBs circular economy recycling companies have to work with battery manufacturers to develop sustainable recycling for all valuable components of batteries.
2. Future studies have to be conducted on the economic viability of graphite recycling and other components of lithium-ion batteries.
3. Future studies are required to analyze the economic viability of sustainable recycling processes on a commercial scale to facilitate the establishment of recycling businesses, improve the electronic waste management policies and legislation of spend LIBs to assist countries achieving an environmentally sustainable framework.
4. Given the rapid advancements in lithium-ion batteries (LIBs) and the limited number of global processing facilities found on a commercial scale to recycle spent LIBs, there is a need to investigate the profitability of the recycling processes used. This analysis should rely on the used recovery technologies, the chemical composition of the valuable materials extracted from spent LIBs, as well as the associated transportation costs, labors costs and the capital costs related to recovery plants.
5. Due to natural reserves are not endless and spent LIBs are not recycled with 90% efficiency, Li natural reserves in the world and new LIBs are being produced with less cobalt content. Furthermore, the adoption of the circular economy approach aimed to minimize waste and ensure the eco-sustainability of discarded products, therefore it is crucial to value the available materials in the next generation of LIBs. Consequently, it is essential to develop innovative recycling processes to effectively establish a profitable value chain for the new generation of LIBs within the context of the circular economy.

- **Recycling Processes Aspects of Spent LIBs:**

1. Although most literature significantly reviews from different perspectives the recycling of spent LIBs, all agree that much more development is needed to achieve sustainable and economic benefits for all types of spent LIBs. Whereas the majority of commercial recycling companies use one or both processes of mechanical recycling, such as hydrometallurgy and pyrometallurgy, focusing on extracting the high precious materials rather than the closed-loop recycling of other component materials, ignoring the technological and environmental gaps of disposed component materials of spent LIBs with lacking the economic thinking of other component materials and the comparison of those recycling processes in the literature.
2. While lithium (Li), manganese (Mn), and graphite are rarely recovered, and the predominant recycling processes have been directed toward the reclamation of cathode materials from spent lithium-ion batteries (LIBs), with comparatively less focusing on other materials such as anodes, electrolytes, and cathodes as well as LIBs designs, only spent LIBs with high content of cobalt (Co) and other precious materials are highly preferred for recycling. Moreover, cobalt and graphite are the main materials in LIBs and are considered mainly sufficient

76. Léonard A, Hantson P, Gerber GB (1995) Mutagenicity, carcinogenicity and teratogenicity of lithium compounds. Mutation Res/Rev Genetic Toxicol 339(3):131–137. https://doi.org/10.1016/0165-1110(95)90007-1
77. MEF (2001) Ministry of Environment and Forests Notification New Delhi, the 16th May, 2001
78. MEF (2019) Ministry Of Environment and Forests Notification New Delhi, the 16th May, 2001, Amended 2019
79. MEFCC (2020) Ministry of Environment, Forests and Climate Change Notification New Delhi, the 20th February, 2020
80. METI (1991) Act on the promotion of effective utilization of resources, 2(48), pp 1–24
81. METI (2015) Summary_Law for the promotion of effective utilization of resources
82. Ma X et al (2018) The recycling of spent power battery: economic benefits and policy suggestions. IOP Conf Ser Earth Environ Sci 159(1). https://doi.org/10.1088/1755-1315/159/1/012017
83. Mahsa B et al (2019) Advances in bioleaching as a sustainable method for metal recovery from e-waste: a review. J Ind Eng Chem 76:75–90. https://doi.org/10.1016/j.jiec.2019.03.047
84. Mancini L et al (2020) Responsible and sustainable sourcing of battery raw materials. JRC Technical Report. https://doi.org/10.2760/562951
85. Mayyas A, Steward D, Mann M (2019a) The case for recycling: Overview and challenges in the material supply chain for automotive li-ion batteries. Sustain Mater Technol 19:e00087. https://doi.org/10.1016/j.susmat.2018.e00087
86. Mayyas A, Steward D, Mann M (2019b) The case for recycling: overview and challenges in the material supply chain for automotive li-ion batteries. Sustain Mater Technol 19:e00087. https://doi.org/10.1016/j.susmat.2018.e00087
87. Mehta R, Templeton DM, O'Brien PJ (2006) Mitochondrial involvement in genetically determined transition metal toxicity. II. Copper toxicity. Chemico-Biological Interact 163(1–2):77–85. https://doi.org/10.1016/j.cbi.2006.05.011
88. Mersky AC et al (2016) Effectiveness of incentives on electric vehicle adoption in Norway. Transp Res Part D Transp Environ 46:56–68. https://doi.org/10.1016/j.trd.2016.03.011
89. Meshram P, Pandey BD, Mankhand TR (2014) Extraction of lithium from primary and secondary sources by pre-treatment, leaching and separation: a comprehensive review. Hydrometallurgy 150:192–208. https://doi.org/10.1016/j.hydromet.2014.10.012
90. Mishima K, Nishimura H (2016) Requirement analysis to promote small-sized E-waste collection from consumers. Waste Manage Res 34(2):122–128. https://doi.org/10.1177/0734242X15615424
91 Morin D, Lips A, Pinches J, Frias C, Norberg A, Forssberg E (2006) BioMinE-Integrated project for the development of biotechnology for metal-bearing materials in Europe. Hydrometallurgy 83(1–4):69–76
92. Mossali E et al (2020) Lithium-ion batteries towards circular economy: a literature review of opportunities and issues of recycling treatments. J Environ Manage, 264. https://doi.org/10.1016/j.jenvman.2020.110500
93. Nayaka GP et al (2019) An environmental friendly attempt to recycle the spent Li-ion battery cathode through organic acid leaching. J Environ Chem Eng 7(1):102854. https://doi.org/10.1016/j.jece.2018.102854
94. Nitta N et al (2015) Li-ion battery materials: present and future. Mater Today 18(5):252–264. https://doi.org/10.1016/j.mattod.2014.10.040
95. Nixon H et al (2009) Understanding preferences for recycling electronic waste in California: the influence of environmental attitudes and beliefs on willingness to pay. Environ Behav 41(1):101–124. https://doi.org/10.1177/0013916507310053
96. Nkulu L et al (2018) Europe PMC Funders Group Sustainability of artisanal mining of cobalt in DR Congo. Nat Sustain 1(9):495–504. https://doi.org/10.1038/s41893-018-0139-4.Sustainability
97. OECD (2014) Case study for OECD project on extended producer responsibility Republic of Korea (May), pp 1–20

98. OECD Green Growth Studies (2012) OECD Green Growth Studies. Director, p 104
99. Olsson L et al (2018) Circular business models for extended ev battery life. Batteries 4(4). https://doi.org/10.3390/batteries4040057
100. Ongondo FO, Williams ID, Cherrett TJ (2011) How are WEEE doing? A global review of the management of electrical and electronic wastes. Waste Manage, 714–730. https://doi.org/10.1016/j.wasman.2010.10.023
101. Padhi AK, Nanjundaswamy KS, Goodenough JB (1997) Phospho-olivines as positive-electrode materials for rechargeable lithium batteries. The Electrochemical Society [Preprint]
102. Parellada JA et al (2001) Patient tracking and follow-up after percutaneous breast biopsy. Appl Radiol 26(2):7–10
103. Peter ALJ, Viraraghavan T (2005) Thallium: a review of public health and environmental concerns. Environ Int 31(4):493–501. https://doi.org/10.1016/j.envint.2004.09.003
104. Qu Y et al (2019) Understanding residents' preferences for e-waste collection in China—a case study of waste mobile phones. J Clean Prod 228:52–62. https://doi.org/10.1016/j.jclepro.2019.04.216
105. Rahman A, Afroz R (2017) Lithium battery recycling management and policy. Int J Energy Technol Policy 13(3):278–291. https://doi.org/10.1504/IJETP.2017.084497
106. Rajaeifar MA et al (2020) Sustainable supply and value chains of electric vehicle batteries. Resour Conserv Recycl 161(June):104905. https://doi.org/10.1016/j.resconrec.2020.104905
107. Rajaeifar MA et al (2019) Emissions from urban bus fleets running on biodiesel blends under real-world operating conditions: implications for designing future case studies. Renew Sustain Energy Rev 111(May):276–292. https://doi.org/10.1016/j.rser.2019.05.004
108. Randell P (2016) Waste lithium-ion battery projections. In: Lithium-ion forums: recycling, transport and warehousing. Randell Environmental Consulting, Woodend, Victoria, pp 1–18
109. Rene ER et al (2021) Electronic waste generation, recycling and resource recovery: Technological perspectives and trends. J Hazard Mater 416:125664. https://doi.org/10.1016/j.jhazmat.2021.125664
110. Richa K (2016) Sustainable management of lithium-ion batteries after use in electric vehicles, p 199
111. Richa K, Babbitt CW, Gaustad G (2017) Eco-efficiency analysis of a lithium-ion battery waste hierarchy inspired by circular economy. J Ind Ecol, 715–730. https://doi.org/10.1111/jiec.12607
112. Ruffino B, Zanetti MC, Marini P (2011) A mechanical pre-treatment process for the valorization of useful fractions from spent batteries. Resour Conserv Recycl 55(3):309–315. https://doi.org/10.1016/j.resconrec.2010.10.002
113. Santana IL et al (2017) Photocatalytic properties of $Co_3O_4/LiCoO_2$ recycled from spent lithium-ion batteries using citric acid as leaching agent. Mater Chem Phys 190(January 2016):38–44. https://doi.org/10.1016/j.matchemphys.2017.01.003.
114. Schmuch R et al (2018) Performance and cost of materials for lithium-based rechargeable automotive batteries', *Nature Energy*, 3(4), pp. 267–278. Available at: https://doi.org/10.1038/s41560-018-0107-2.
115. Shi Y, Chen G, Chen Z (2018) Effective regeneration of $LiCoO_2$ from spent lithium-ion batteries: a direct approach towards high-performance active particles. Green Chem 20(4):851–862. https://doi.org/10.1039/c7gc02831h
116. Simate GS, Ndlovu S, Walubita LF (2010) The fungal and chemolithotrophic leaching of nickel laterites—challenges and opportunities. Hydrometallurgy 103(1–4):150–157. https://doi.org/10.1016/j.hydromet.2010.03.012
117. Sommerfeld M et al (2020) A combined pyro-and hydrometallurgical approach to recycle pyrolyzed lithium-ion battery black mass part 1: production of lithium concentrates in an electric arc furnace. Metals 10(8):1–27. https://doi.org/10.3390/met10081069
118. Sommerville R et al (2021) A qualitative assessment of lithium ion battery recycling processes. Resour Conserv Recycl 165(October 2020):105219. https://doi.org/10.1016/j.resconrec.2020.105219

119. Sonoc A, Jeswiet J (2014) A review of lithium supply and demand and a preliminary investigation of a room temperature method to recycle lithium ion batteries to recover lithium and other materials. Procedia CIRP 15:289–293. https://doi.org/10.1016/j.procir.2014.06.006
120. Sonoc A, Jeswiet J, Soo VK (2015) Opportunities to improve recycling of automotive lithium ion batteries. Procedia CIRP 29:752–757. https://doi.org/10.1016/j.procir.2015.02.039
121. Sun M et al (2015) Consumer behavior and perspectives concerning spent household battery collection and recycling in China: a case study. J Clean Prod 107:775–785. https://doi.org/10.1016/j.jclepro.2015.05.081
122. Swain B (2017) Recovery and recycling of lithium: a review. Separ Purif Technol 172:388–403. https://doi.org/10.1016/j.seppur.2016.08.031
123. Tabassum H et al (2018) A universal strategy for hollow metal oxide nanoparticles encapsulated into B/N co-doped graphitic nanotubes as high-performance lithium-ion battery anodes. Adv Mater 30(8):1–7. https://doi.org/10.1002/adma.201705441
124. Tan Q et al (2018) Rethinking residential consumers' behavior in discarding obsolete mobile phones in China. J Clean Prod 195:1228–1236. https://doi.org/10.1016/j.jclepro.2018.05.244
125. Tang Y et al (2018) Recycling mechanisms and policy suggestions for spent electric vehicles' power battery—case of Beijing. J Clean Prod 186(2018):388–406. https://doi.org/10.1016/j.jclepro.2018.03.043
126. Tasaki T (2014) The act on the promotion of effective utilization of resources and the treatment of specified resources-recycled products stipulated in the act
127. Torres D et al (2016) Sustainable management of waste electrical and electronic equipment in Latin America. ITUPublications, pp 1–66
128. Trading Economic (2021) Cobalt 2010–2021 Data | 2022–2023 Forecast | Price | Quote | Chart | Historical. Available at: https://tradingeconomics.com/commodity/cobalt
129. Tran MK et al (2019) Deep eutectic solvents for cathode recycling of Li-ion batteries. Nat Energy 4(4):339–345. https://doi.org/10.1038/s41560-019-0368-4
130. Treffer F (2018) Lithium-ion batteries: basics and applications. In: Lithium-ion batteries: basics and applications, pp 1–413. https://doi.org/10.1007/978-3-662-53071-9
131. Tsiropoulos I, Tarvydas D, Lebedeva N (2018) Li-ion batteries for mobility and stationary storage applications. JRC Science for Policy Report, EU Commission, Publications Office of the European Union. https://doi.org/10.2760/87175
132. Turner JM (2015) Following the Pb: an envirotechnical approach to lead-acid batteries in the United States. Environ History 20(1):29–56. https://doi.org/10.1093/envhis/emu128
133. Wang X et al (2014b) Economies of scale for future lithium-ion battery recycling infrastructure. Resour Conserv Recycl 83:53–62. https://doi.org/10.1016/j.resconrec.2013.11.009
134. Wang X et al (2014a) Economies of scale for future lithium-ion battery recycling infrastructure. Resour Conserv Recycl 83:53–62. https://doi.org/10.1016/j.resconrec.2013.11.009
135. Wang Z, Guo D, Wang X (2016) Determinants of residents' e-waste recycling behaviour intentions: evidence from China. J Clean Prod 137(February 2009):850–860. https://doi.org/10.1016/j.jclepro.2016.07.155
136. Whittingham MS (2004) Lithium batteries and cathode materials. Chem Rev 104(10):4271–4301. https://doi.org/10.1021/cr020731c
137. Winslow KM, Laux SJ, Townsend TG (2018c) A review on the growing concern and potential management strategies of waste lithium-ion batteries. Resour Conserv Recycl 129(November 2017):263–277. https://doi.org/10.1016/j.resconrec.2017.11.001
138. Winslow KM, Laux SJ, Townsend TG (2018b) A review on the growing concern and potential management strategies of waste lithium-ion batteries. Resour Conserv Recycl 129(October 2017):263–277. https://doi.org/10.1016/j.resconrec.2017.11.001
139. Winslow KM, Laux SJ, Townsend TG (2018a) A review on the growing concern and potential management strategies of waste lithium-ion batteries. Resour Conserv Recycl 129(October 2017):263–277. https://doi.org/10.1016/j.resconrec.2017.11.001
140. Xin Y et al (2016b) Bioleaching of valuable metals Li, Co, Ni and Mn from spent electric vehicle Li-ion batteries for the purpose of recovery. J Clean Prod 116:249–258. https://doi.org/10.1016/j.jclepro.2016.01.001

141. Xin Y et al (2016a) Bioleaching of valuable metals Li, Co, Ni and Mn from spent electric vehicle Li-ion batteries for the purpose of recovery. J Clean Prod 116:249–258. https://doi.org/10.1016/j.jclepro.2016.01.001
142. Xu J et al (2008) A review of processes and technologies for the recycling of lithium-ion secondary batteries. J Power Sources 177(2):512–527. https://doi.org/10.1016/j.jpowsour.2007.11.074
143. Yang Y et al (2020) On the sustainability of lithium ion battery industry—a review and perspective. Energy Storage Mater [Preprint]. https://doi.org/10.1016/j.ensm.2020.12.019
144. Yang Y, Xu S, He Y (2017) Lithium recycling and cathode material regeneration from acid leach liquor of spent lithium-ion battery via facile co-extraction and co-precipitation processes. Waste Manage 64:219–227. https://doi.org/10.1016/j.wasman.2017.03.018
145. Yin J, Gao Y, Xu H (2014) Survey and analysis of consumers' behaviour of waste mobile phone recycling in China. J Clean Prod 65:517–525. https://doi.org/10.1016/j.jclepro.2013.10.006
146. Ylä-Mella J, Keiski RL, Pongrácz E (2015) Electronic waste recovery in Finland: consumers' perceptions towards recycling and re-use of mobile phones. Waste Manag 45(2015):374–384. https://doi.org/10.1016/j.wasman.2015.02.031
147. Yu J et al (2018) A promising physical method for recovery of $LiCoO_2$ and graphite from spent lithium-ion batteries: grinding flotation. Separ Purif Technol 190(June 2017):45–52. https://doi.org/10.1016/j.seppur.2017.08.049
148. Yun L et al (2018) Metallurgical and mechanical methods for recycling of lithium-ion battery pack for electric vehicles. Resour Conserv Recycl 136(January):198–208. https://doi.org/10.1016/j.resconrec.2018.04.025
149. Zabala A (2019) Illegal electronic waste recycling trends. Nat Sustain 2(5):353–354. https://doi.org/10.1038/s41893-019-0289-z
150. Zeng X, Li J, Liu L (2015) Solving spent lithium-ion battery problems in China: opportunities and challenges. Renew Sustain Energy Rev 52:1759–1767. https://doi.org/10.1016/j.rser.2015.08.014
151. Zero Waste SA (2017) Zero Waste SA revocation regulations 2017 Part 1—preliminary, (6), p 2017
152. Zhao J et al (2019) Extraction of Co and Li_2CO_3 from cathode materials of spent lithium-ion batteries through a combined acid-leaching and electro-deoxidation approach. J Hazard Mater 379(May):120817. https://doi.org/10.1016/j.jhazmat.2019.120817
153. Zheng M et al (2018) Hierarchically nanostructured transition metal oxides for lithium-ion batteries. Adv Sci. https://doi.org/10.1002/advs.201700592
154. Zhou Y et al (2015) Plug-in electric vehicle market penetration and incentives: a global review. Mitigation Adaptation Strat Global Change 20(5):777–795. https://doi.org/10.1007/s11027-014-9611-2
155. Zhou L, Xu Z (2012) Response to waste electrical and electronic equipments in China: legislation, recycling system, and advanced integrated process. Environ Sci Technol [Preprint]. Available at: https://doi.org/10.1021/es203771m

Food Systems Within a Circular Economy Framework

İdris Akkuzu, Seda H. Bostancı, Cristina Raluca Gh. Popescu, and Gheorghe N. Popescu

Abstract The circular economy is a system that combines long-term use, maintenance, repair and waste collection to minimise environmental damage in the production and consumption processes of products. One of the various dimensions of circular economy is circular food systems. As a result of the production cycle in food systems, the problem of food waste arises because the recycling product obtained after the food becomes waste cannot be used for food for people. Waste is not only a key input in the circular economy, but also a policy issue. Excessive amounts of household waste, including food waste, are generated when policy makers favour competitive markets and value consumption over production. When the issue is approached from this perspective, it is important to investigate the gains of food systems to the circular economy as well as the background of waste policies. It is thought that neoliberal policies have negative effects on the increase in hunger and extreme poverty problems together with overconsumption and waste worldwide. In this study, neoliberal food policies are analysed with this approach and policy recommendations on circular food systems are tried to be developed. Since the study is a conceptual analysis, a literature review has been conducted and the deficiencies and inconsistencies in the way the concepts are handled in the literature have been pointed out. The reason for these shortcomings and inconsistencies is that in the analyses of circular economy and circular food systems, the commodifying and thus marketising capability of neoliberal economic and political processes on things, inventions and systems that

İ. Akkuzu · S. H. Bostancı (✉)
Faculty of Economics and Administrative Sciences, Department of Political Science and Public Administration, Tekirdağ Namık Kemal University, Tekirdağ, Türkiye
e-mail: shbostanci@nku.edu.tr

İ. Akkuzu
e-mail: idrisakkuzu@gmail.com

C. R. Gh. Popescu · G. N. Popescu
The Bucharest University of Economic Studies, Bucharest, Romania
e-mail: cristina.popescu@man.ase.ro

G. N. Popescu
e-mail: gheorghe.popescu@cig.ase.ro

A. Cavalcante da Silva Batalhão and P. Singh (eds.), *Blending Approaches Towards Sustainability Focus*, Earth and Environmental Sciences Library,
https://doi.org/10.1007/978-3-031-93518-3_5

are not yet subject to the market is overlooked. In this respect, the study approaches the relevant literature with a critical review method.

Keywords Political economy · Circular economy · Neoliberal food policies · Circular food systems · Alternative food networks

1 Introduction

Food policies are among the connecting areas of food systems and circular economy. These systems are affected by economic, natural and social crises around the world, and poverty-related hunger and malnutrition problems arise in fragile economies. Problems of access to food are among the main issues awaiting solutions in the field of social justice. Food justice debates around the world are closely linked to economic crises, climate change, migration and war. According to International Food Policy Research Institute (IFPRI) [28], ideal food systems should primarily provide nutrition, access to safe and healthy food, be productive at affordable prices, be resilient to climate change, and be environmentally sustainable.

Within the concepts of circularity, there are policy and implementation tools for recycling and reuse. Within this system, policy makers are expected to develop more appropriate means of use and to design a circularity sharing system based on social solidarity, instead of uncontrolled and competitive production systems. Recycling opportunities are advancing with field-related technology and innovative approaches. In order to make food systems circular, it is important for citizens to gain awareness of recycling food waste, especially in the household waste process, as well as in agricultural and production processes. Documentaries, public service announcements and social media applications are used as a policy tool in various countries to raise awareness among citizens.

Orthodox economics, in layman's terms, has an approach that treats the production processes of things as a linear system and places at the centre of the discipline an abstract market mechanism driven by the maximisation objectives of individuals who do not worry about the before and after processes. This approach, again in the most corny terms, accepts that the market can produce inefficient outcomes, provided that a solution is possible, and argues that the mechanisms to compensate for these outcomes can be produced through economic policies, again by internalising them into the market system. In this respect, for example, it is argued that problems such as environmental damage, externalities, lack of access to resources and markets that may be caused by free market processes do not need to be sought outside the concept sets offered by orthodox economics, but can be expressed in a formal way by incorporating the concepts of disciplines such as political science and social policy into the general operating principles of economics. In fact, this approach is valid not only for ecology and food issues, which are the subject of this study, but also for all social issues. For this domination of the discipline of economics over other disciplines, the concept of Economic Imperialism can be used, borrowing from Ben

Fine, at the expense of leaving aside the explanation of its historical context and how it evolved here, which is beyond the scope of this study.

The relationship between circular economy and economics presents a dual view in this respect. First, while circular economy appears to be a challenge to the linear functioning of orthodox economics, on the other hand, it has the potential to become a useful apparatus that economics can internalise and formalise. In this respect, it has become a field of study that has recently attracted the attention of both orthodox economists and heterodox approaches. However, what needs to be emphasised here is the necessity of placing the relationship between economics and circular economy in a historical context. This is neoliberalism itself for the last 40 years, in addition to the globalisation and centralisation of capital, neoliberalism, with an unrelenting appetite, has been trying to confine areas such as nature, public goods, social values, culture and science, which were previously largely left outside the market, to an economistic sphere. It is this aspect of the issue that is critical for the circular economy Because neoliberal political economy tends to marketise the circular economy by turning it into a commodity and condemning it to market processes by extracting rents from it For this reason, circular economy should not be discussed only in terms of a formal mode of explanation, but should be discussed in terms of policies and inter-class relations and in the context of the global division of labour.

From a food policy perspective, the issue of food has particularly critical links with social justice. Indeed, as Marsden and Morley [38] argue, for fair distribution, public health and healthy ecosystems worldwide, food systems should be planned with social values in min. One of the research areas related to this issue is working-class environmentalism. Researchers interested in this topic point out that the destruction of nature and the creation of toxic living and working environments for capital accumulation and excessive profits constitute a common area of struggle for grassroots organisations, workers and environmentalists [4, 43]. These toxic environments threaten public health by creating environments conducive to environmental pollution, the spread of infectious diseases. The most affected by these conditions are the extreme poor. The aspects of neoliberal food policies that increase consumption in all areas cause a significant amount of food to be wasted. Food policy based approaches have been insufficient to prevent hunger and malnutrition worldwide. The fact that the circular economy is presented as a solution to food waste makes it difficult to take more radical measures against waste. Thus, neoliberal processes in which food consumption is encouraged gain continuity. While the least developed countries, defined as LCDs around the world, struggle with hunger, they also lack access to clean water resources. Problems related to clean water resources may cause civil wars in these countries [6, 21]. The use of clean water resources in food production processes also exacerbates the water problem, which is considered to be the main problem worldwide, at the same time as food waste. Along with fair distribution policies in these areas, sanctions especially for production processes should be increased. Alternative food networks have started to be established in this field, which combines environmental problems and public health worldwide. These alternative food networks include strategies, methods and techniques for the development of local food systems. Alternative food networks linked to circular food systems include specialised forms of farm-to-direct retailing, community-supported agriculture, fair trade and organic farming [12, 22, 42].

While examining food systems in the circular economy, the study questions the problem of food waste, with the assumption that it will become a more critical issue for the world of the future. The literature review identifies that the root of the problem is linked to neoliberal food policies and provides a conceptual analysis including a system critique. The study discusses social justice-based policy proposals for food systems, including the concept of alternative food networks, and compares different approaches to treating food systems as part of the circular economy. Here, the possibilities of transition from the production and consumption cycle to fair and balanced distribution processes are discussed. In particular, approaches to the development of local food systems through the concepts of food sovereignty and food justice are emphasised. In this context, the main approach of the study is to establish the link between circular economy and responsible economy by focusing on alternative food systems instead of instrumentalising circular economy for the continuity of global food systems.

2 Defining Circular Economy

The concept of circular economy began to be discussed in the 1990s. However, Zucchella and Previtali [62] mention that the foundations of this approach were laid in the 1970s with the concepts of "lifecycle assessment" and "cradle-to-cradle". The issues related to the concept are environmental economy, industrial economy and ecological economy [34]. When circular economy is associated more broadly, it is seen that it has links with the concepts of industrial ecology, cradle-cradle design, ecological economics, resilience science, natural capitalism, and cleaner production [13, 33]. These contents show that the concept has controversial aspects that sometimes contradict each other. There are key issues that different schools of thought focus on in the circular economy. Examples of these include approaches to minimizing resource and waste extraction, economic growth potential, and reducing negative environmental impacts [61].

The circular economy creates various models. Some of these models extend the product's life through remanufacturing, repair, and improvements, while another approach uses recycling techniques to transform old products into new resources [53]. For economists working on gross domestic product (GDP), extending the use of products and developing systems for reuse seem contradictory to the economic growth approach. In addition, considering the sales and advertisements of recycled products, it is an appropriate approach to express circularity approaches as natural capitalism.

"Strategies to support local food produce a policy based on social justice as it creates a circular economy that enables the production of shareable and affordable food for all" [60]. Circularizing local food systems not only increases local competitiveness and economic development, but also provides citizens with social solidarity and the ability to produce their own systems.

Corvellec et al. [13] criticized the circular economy concept as being normative, its over-emphasis on positivity, its consensual win–win policies. Another issue that can be criticized regarding the circular economy is the uncertainty regarding the future of raw materials. Geng et al. [20] mention that recycling a material that poses a risk to sustainability, such as plastic, may create the risk of producing more plastic. The argued that when lower prices increase demand, more plastic will be needed for recycling.

Critical studies on the circular economy are limited. In one of these limited studies, Niskanen et al. [45] analysed the attitudes of public officials, opinion leaders and political parties towards the circular economy in press materials on Sweden. They found that in Sweden, ideas about the circular economy are used as a new expression of ecological modernisation. They also pointed out that it could have been used by resilient capital to contain an economically destructive process. They conclude that circular economy ideas need to be more critically analysed in the field of political economy.

3 Circular Food Systems

Food systems as a chain process include production, processing, distribution and consumption stages. It is appropriate to associate this process with the economic dimension of import and export, with the health dimension of food security, and with the social dimension of food justice. In simple terms, food systems can be defined as the totality of activities that provide people with access to food [52]. According to IFPRI [28], food systems are the sum of economic and actor-based interactions of the food chain, from livestock, crops, fish, and other agricultural commodities as food inputs to their transportation, processing, preparation, wholesale, consumption, waste generation, recycling, and disposal. The primary function of food systems is food supply. Local extreme weather conditions or strikes, global market price anomalies, sudden migration or pandemics affecting food demand, and conflicts in trade policies can all lead to supply disruptions [25]. In these circumstances, food stocks are strategic issues for countries.

The agricultural sector, which is the basis of food, is very vulnerable to the negative consequences of climate change [15]. The overproduction pressure created by multinational agribusiness companies in countries that support neoliberal agricultural policies seriously threatens the sustainable future of these regions. Enabling environments for human capital and innovation support sustainable food systems based on technological innovation [51]. Sustainable food systems are expected to be resilient, sustainable, health promoting, diverse, fair, economically balanced and transparent [26].

Circular food systems replace diverting food waste from landfills or capture with recycling, thus increasing soil fertility. Methods such as reducing energy and water use in food production, anaerobic digestion and composting using food waste treatment technologies, and stabilizing waste are used in circular food systems [57]. The basis of circularity is the combined use of policies, strategies and technologies to achieve zero waste or to minimise waste. From this perspective, circularity can be expressed as a set of approaches and measures regarding waste. In this respect, in order to understand a circular food system, it would be appropriate to define the concepts of food waste, food loss and surplus food. Food loss was defined by Teigiserova et al. [56] as the decrease in the quality and quantity of edible food in general before consumption, with post-harvest losses, taking into account the relevant literature and FAO (Food and Agriculture Organization of the United Nations) and IFPR approaches. When looking at the process from a circular economy perspective, decisions taken on evaluating the products left in the field together with the social assistance policies of countries can greatly reduce losses in this regard. "Food waste is a subset of food loss, and consists of material intended for human consumption that is not consumed" [32]. From this perspective, food loss is the responsibility of the producer and food waste is the responsibility of the consumer. Food waste contributes to circularity by encouraging citizens to recycle their food waste and reuse this waste with appropriate technologies, primarily through the encouragement of local governments or volunteers. Surplus food can be defined as food that is still edible but is rejected by producers or retailers for reasons such as aesthetic appearance or lack of demand [17]. When food with these properties cannot be used quickly, it becomes food waste. Surplus food can be defined as uncontrolled and unusable food that is the result of excessive consumption brought about by the neoliberal order. This process makes the circular economy controversial. Here, systems such as open buffet services in tourism facilities and restaurants are shown as a problem area that can be controlled with the circular economy. However, while the circularity of the food system provides a solution to the waste problem, the loss of productive food continues to be large. Thus, while food is wasted at serious rates due to certain economic and political choices, hunger and malnutrition problems continue in certain parts of the world.

Food redistribution policies are an area where governments, local governments and NGO's seek different solutions. These policies have content related to food justice. Food justice is an area of advocacy regarding food sovereignty and food access. Food sovereignty is the right of a nation, a society, to produce and define its own food and agricultural systems [24, 49]. Neoliberal policies violate food sovereignty, especially in countries defined as LCDs. Throughout history, colonialism has seriously threatened the ecological and agricultural sustainability of LCDs by exploiting their agricultural products, lands and forest areas. "Food justice is among the main indicators of countries for understanding the democracy approach, economic level, conflict situations such as civil war conditions and migration and also political approaches such as social justice" [60]. From this perspective, although the circular economy has various tools for a circular food system, it becomes clear that the issue needs to be addressed with more socially just policies in food systems due to the deep insight into the food issue and its being at the root of injustices in the world. With this approach, it will be examined how local food systems and alternative food networks will contribute to the process in a way that supports food sovereignty and food justice, taking into account critical approaches based on literature on neoliberal food policies.

4 Criticizing Circular Food Systems by Neoliberal Food Policies

According to the World Research Institute, there is a 56% gap between the amount of food that will be needed to feed the world population in 2050 and the amount of food produced today [59]. This estimate of global food demand pressure, especially when the unstable economic and political conditions of the global south and the economic behaviour of northern-based global corporations that exclude social sensitivity are taken into account, constitutes a risk factor for food production and distribution, and poses a threat to access to healthy food for a significant part of the world population. At this stage, the concept of food security, which takes into account the aforementioned risks and is formed with a social sensitivity accordingly, offers a potential for the solution of the global food problem, despite the difficulty of its implementation due to the conflicts of interest between the parties involved. The main issue that would undermine this potential is the effort to bring together different social strata in the centre of market sovereignty. Here, the concept of food security remains incomplete and inadequate as it is far from being a mechanism to deliver the rights of the southern countries, which are positioned in unequal exchange relations in the global political economy, and the disadvantaged groups within the borders of each country, which are also within the extension of the same unequal exchange relations.

As Clapp and Moseley [8] argue that the policy responses to crises in food systems over the past 70 years have promoted a food system based on the principles of specialization, industrial production and global trade. While this system has developed multinational companies, it has created various positive effects in terms of ensuring peace conditions thanks to global trade and supply chains between countries. However, the decrease in self-sufficiency of countries and the increase of the market share of imported products cause deep global poverty as well as the impoverishment of the nations. Mainstream economics is seen as providing the main arguments in politics, presenting an apolitical technocracy in terms of policies to increase GDP. However, the injustices, environmental problems and social events show that this economic approach has lost its appeal [40]. At this stage, Kallis' [31] discourses of degrowth take their place in the relationship between economy and circularity. Criticizing the ideology behind the dogma of economic growth, degrowth is an approach that states that it is possible to move to a different political-economic system with a smaller flow of resources through radical political and economic restructuring [31].

In particular, it is seen as a policy that can produce a degrowth solution to the surplus food problem. Food justice activists address some solution areas against neoliberal food policies there are, organizing labor, cooperative ownership, and pushing to outlaw risky technologies [1].

When food justice movements are examined, similar to environmental justice and other social justice pursuits, the connections between gender, race and class inequalities across countries and global scales come to the fore. There are neoliberal limitations to market-based solutions to the inequalities and problems related to race and class that shape the production, distribution and consumption of food [37].

Neoliberal policies, along with social injustice and deep poverty, have caused serious criticism and social movements from environmental scolars and environmental activists, as they produce investments and policies that create serious environmental problems. The connection between environmental and food issues lies in the subject area of industrial economics. Martínez-Alier [39] states that the industrial economy is entropic, not circular, and produces polluting waste because it requires new energy and material resources. These policies create ecological problems and pollute clean water resources as well as food. As a result, she states that ecological distribution conflicts arise over environmental justice issues such as food justice.

Neoliberal food security policies have failed to prevent hunger and reduce inequalities [19]. The reason for this is the problems in managing the process with large-scale industrial economy-based and multinational companies that are not sensitive to local values and sustainability. As an alternative, Friesen [19] proposes the concept of food sovereignty, discussed in previous sections. There are various definitions of the circular economy, some of which link the subject to the mainstream industrial economy. From this perspective, the circular economy is shown as a tool to reduce the problems in terms of neoliberal food policies. At the same time, the flexibility of the circular economy to adapt to more egalitarian social policies creates an opportunity to associate this process with local food systems, alternative food networks, and public food procurement programs. These concepts will be discussed in detail in the policy recommendations section.

5 The Policy Recommendations for Circular Food Systems

Producing sufficient and healthy food for the world's population, developing food policies that are prepared for complex global changes, and minimizing environmental problems caused by food production are issues that policy makers in food systems are seeking solutions for [52]. Innovative policies have been sought to ensure that food systems are less affected by the economic uncertainties experienced today, sudden global crises such as migration, pandemics and war, and predictable crises such as climate change. The main objectives of these policies are circularity, food security and the development of local food systems that enable countries to be self-sufficient in terms of food. Although the circular economy has criticized areas, this approach helps environmental continuity and biodiversity by recycling. Stahel [53] states that it would be appropriate for governments and policy makers to promote circular economy in industry, including through the option of taxation.

Food security, as access to clean and healthy food, is among the basic human rights. Food security is one of the basic criteria for ensuring social justice. In the mid-1980s, in response to the decline of state support for agriculture and food imports in Central America, the "food sovereignty movement" emerged and the concept was expanded to mean the right of each country to define its own agricultural policy in the interest of its nation and in harmony with peasant and indigenous organizations, ensuring their real participation [7]. "The global movement towards food sovereignty has been seen as an attempt to reclaim rights and participation in the food system and challenge corporate food regimes" [9]. Countries that can develop decisive policies on food sovereignty also have an advantage in implementing welfare state policies and preventing progress in food justice from being lost. However, the state's transformation of this redistribution process into a social assistance regime creates a system problem. Presenting a fundamental right as a political tool in elections is contrary to the concept of a social state. From this perspective, public food procurement (PFP) programs, as a new approach, contribute to public health and quality of life. This program supports local and healthy food, encourages food to be purchased from family farmers, women, youth and other vulnerable groups, and encourages agricultural products to have characteristics that contribute to biodiversity and promote sustainable agriculture [55]. While PFP developed by the UN to support sustainability is spreading worldwide, applications in this field are related to issues such as ensuring biodiversity, providing food security for vulnerable groups, revitalizing rural economies, supporting family farming and developing sustainability approaches in food production [14]. However, governments that adopt neoliberal policies do not support PFP programs or provide symbolic support.

Food activism is a social movement concerned with racial, social, gender, and economic injustice. This includes struggles against Genetically Modified Organisms (GMOs) and pesticides, critiques of consumerism and dependence on private property in the food process, and efforts to improve wages and working conditions for workers in the food system [2]. The importance of food supply and cooking in household chores makes this an area where women have more responsibility among care duties. In terms of its relevance to the issue of care, the fields of feminist care theories, feminist food sovereignty, and feminist political ecology also join the literature of food activism [36]. Thus, the field of food takes its place within gender and labour studies.

In addition to the circularity of fair, egalitarian, socially just food systems, their locality is important. Responsive economy, is a concept that is similar to the circular economy in terms of considering environmental values but has different aspects, contributes to local and sustainable food policies. Responsible economy is an environmental and social justice-based approach where basic environmental resources (such as drinking water and quality soil) are given more importance than in free market economies, and common goods and common assets are taken into consideration [44].

Developing local food systems (LFS) is the most appropriate strategic decision for the economic development and independence of a society, as well as for well-being conditions [18]. From this perspective, local food systems form the foundations of a society consisting of well-nourished, healthy individuals. Local food systems are becoming dependent on global food production, logistics and trade [35]. This process, along with the loss of forest areas, causes the traditional agricultural culture to be forgotten. Agricultural activities that take environmental values into consideration make a significant contribution to the development of healthy and sustainable food systems. Climate friendly food and food footprint can be defined as indicators for sustainable food systems. Among all food products, agricultural production causes the most greenhouse gas emissions, which are among the main causes of climate change. From this perspective, agricultural production policies should be developed for climate-friendly food production [47]. While climate problems are experienced with the conversion of forest areas to agricultural areas, climate problems affect food quality and soil fertility. As the world population increases under these conditions, climate and population data are seen as vitally important in developing international policies on food systems. Globally, there is a risk of acute food shortages in various regions due to fluctuations in yields due to climate change [54]. Migration from these regions sometimes leaves women, children and the elderly behind. These conditions reveal the need to develop new food policy tools for such vulnerable communities.

Similar to other ecological footprints, the food footprint also has criteria related to the sustainability of the environment and public health. The Land Food Footprint (LFF) provides an opportunity to assess the extent to which the local food production system supports the food consumption levels of the local population by assessing land requirements and land demands [35]. Another dimension of the food footprint is the consumer food footprint. Here, the effects of the food consumed by society on the environment, the amount of food consumed, its characteristics and how it is produced, and what ecological footprints it creates while being produced are examined [23]. Raising awareness among individuals about their food footprint can encourage them to prioritize food recycling and support initiatives that contribute to circular food systems.

Civil society-based alternative food networks (AFNs) are an important tool in developing local food systems. AFNs encompasses a number of interconnected initiatives with the aim of creating an alternative to the globalised industrial food production and system [5]. Alternative food systems are also called traditional food systems [46]. Environments conducive to community-supported agriculture, such as urban gardens, create new connections between farmers and consumers in urban areas or in rural areas with new residents [29]. While creating alternative food networks, the development of citizen-based food initiatives provides a strategic approach to the spread of fair and democratic circular food systems. Ayalp [3] grouped citizen-based food initiatives as food communities/community-supported agriculture groups, farm to school, peasant markets and ecological markets, consumer cooperatives, buyer's clubs, urban gardens, and various organizations based on individual initiative. Although these approaches seem innovative and sustainable, they also have features that can cause various problems in practice. With this approach, it would be appropriate to provide brief information about the approaches grouped as citizen-based food initiatives and to mention the challenging parts in their content.

Food communities and community-based food systems (CFS) include various strategies for communities to access food and receive agricultural support to improve local food. These strategies include making local and fresh food more accessible, promoting local food through policy interventions, and training for nutrition, food and agricultural entrepreneurship [10].

Community supported agriculture groups (CSA) are a collective approach that aims to establish direct food relationships between farm members or shareholders and farmers [11]. As part of this system, consumers establish direct relationships with local farmers, and with this approach, gains such as revitalizing local agricultural economies, increasing social food security, and protecting agricultural lands are achieved [48]. The risks of this process include the systematic deterioration of agricultural areas, especially as some entrepreneurs who migrate from cities to rural areas, join these communities, and farm on small-scale lands quickly give up on this process and return to cities.

Farm to school programs, especially implemented in developed countries such as the US, aim to help students observe the experiences of farmers and producers and gain knowledge about food systems, thereby changing their attitudes in this area. These programs provide partnerships between schools and local farms, providing local and healthy food, improving student nutrition, and supporting small and medium-sized local farmers [30]. One of the important programs to ensure that children of school age have equal and fair access to food is the free school meals application [50]. Free school meals programs are supported by governments or local authorities. They are seen as part of equal opportunities in education.

Shopping at farmers' markets creates personal interactions that are expressed as a return to the local [58]. Farmers' markets and ecological markets facilitate access to nature and natural food for communities that are sensitive to ecological values and adapt to healthy nutrition conditions. The biggest problem of these systems is that organic production is more expensive than other products due to the supply–demand relationship. Thus, ecological markets are turning into luxury consumption areas.

Food cooperatives and consumer cooperatives are also among the AFNs models. Consumer initiative cooperatives can be expressed as initiatives that aim to create a dialogue environment for farmers to reduce chemical use by raising awareness about the nature and seasonality of farming [46]. The biggest problem in these initiatives that create interest from cities to rural areas is that people who do not have experienced in the agricultural sector and have no previous experience in this field are generally not able to contribute sufficiently to the dynamics of the process.

Food buyer's club can be defined as a group of individuals who form a community, use their collective purchasing power to buy products (food) in bulk quantities and sell products that are hard to find and have lower unit prices than traditional grocery stores, in a collaborative manner based on the group's benefit [27]. This system may or may not be applicable depending on the laws and administrative systems of the countries. These initiatives become quite controversial, especially in times of pandemics and civil unrest and conflict. The fact that high-income people who are concerned about the future of the world stockpile food, and sometimes countries stockpile similar food, increases injustice.

Urban gardens, unlike hobby gardens, are areas where training is provided and where urban agriculture is possible, sometimes through individual ownership, sometimes with cooperative or municipal support [16, 41]. These areas, especially in areas where local initiatives allow, are divided into small parcels, allowing different people to cultivate the areas. At the same time, interaction and a sense of collectivity among citizens develop in these areas. The increase in such areas allows cities to develop with healthier, cleaner air and the development of qualified green areas. In these aspects, these initiatives can be defined as an AFN that supports both ecological and social life. In various countries, governments are implementing zoning approaches to convert urban gardens into residential areas. Such situations pave the way for citizens to defend these areas and thus develop conscious urban and environmental activism.

6 Conclusion

The study addressed the relationship among food systems, the circular economy and the critique of neoliberal food policies. While developing the content, various connections were established between current concepts. When viewed from the perspective of food policies, it is striking that there is a great diversity of concepts. However, it is seen that the main emphasis on the food issue must be on hunger area but also this has the risk of being located to the background in this confusion of concepts. When the process is approached from this perspective, it was thought that a critique of the neoliberal order was necessary in the context of food policy.

In the study, the negative effects of multinational companies that do not consider local values in the global market on production and the neoliberal policies on the failure to solve hunger problems in LCDs are quoted from the relevant literature. At this stage, when the issues of food loss, food waste, and surplus food are examined, the deadlocks created by current policies around the world are more clearly understood when millions of people living in hunger and malnutrition conditions are considered.

Circular economy is accepted as a sustainable food policy approach, especially because it can produce various solutions regarding food safety and preventing food waste. While this is appropriate, the circularity embedded in neoliberal food policies poses a risk to the continuation of the current order. Alternatively, concepts such as responsive economy and degrowth, which consider environmental values in a different context, may offer opportunities for the development of different approaches to food systems. Since the economic-political content of each topic is quite in-depth, the study was able to briefly explain the relationships of different concepts with food systems, circular economy and criticizing neoliberal food policies. In this respect, since the administrative dynamics and political structures of each country are different, the combinations of these contents can be addressed at the country and regional scales in future studies. Issues such as labour, gender, poverty and inequalities in the food field were not briefly covered in the study, and since the connections of the process with the issue of migration are quite complex, the issue of migration was left out of the scope.

The solution suggestions in the study are generally focused on developing local food systems. In this way, an environment can be provided that will enable countries to develop their food sovereignty and become resilient in their regions with their self-sufficient food systems. It is seen that developed countries are responsible to other countries in terms of food access and food security and that sustainable development goals are inadequate in this area. Public food procurement (PFP) programs also draw attention as an important social policy approach within the scope of the study. The relationship of all these areas with civil society organizations and social movements is addressed within the research area of food justice and food activism.

Here, a few initiatives are discussed under the title of AFN and brief definitions of those discussed are provided. It is seen that alternative initiatives are as effective as activism in the critique of neoliberal food policies. In particular, the support of citizen-based food initiatives by local governments together with civil society organizations can create effects that can transform public policies in the field of food. The most important risk of these initiatives is that initiatives such as urban gardens and ecological markets will develop a system based on competition in land ownership relations, and ecological market products will become expensive and luxury consumer products in the market. At the same time, the rapid increase in the number of AFNs, such as farmers' markets, especially for farmers, and the lack of sufficient support after a certain period, reduces the interest of farmers in this area, causing agricultural land to be sold and converted into residential areas. Under these conditions, it is necessary to produce policies that will ensure the sustainability of AFN initiatives and, when conditions are suitable, to unite these initiatives under a collective roof. Academic studies covering a long time are needed in this field.

As a result of the connections to be established between all these concepts by scolars and civil society organizations, policies that support the development of citizen-based food initiatives that emphasize local values and are open to citizen interaction should be produced and disseminated by putting pressure on governments and local administrations in this area. Countries that can achieve this can provide self-sufficiency and social justice in cases of sudden food crises.

References

1. Alkon AH (2014) Food justice and the challenge to neoliberalism. Gastronomica J Food Culture 14(2):27–40
2. Alkon A, Guthman J (eds) (2017) The new food activism: opposition, cooperation, and collective action. University of California Press
3. Ayalp EK (2021) Alternatif gıda ağları ve Türkiye'de yurttaş temelli gıda inisiyatifleri. İdealkent 12(33):965–1005
4. Barca S (2012) On working-class environmentalism: a historical and transnational overview. Interface J About Soc Movements 4(2):61–80
5. Bilewicz A, Spiewak R (2015) Enclaves of activism and taste: consumer cooperatives in Poland as alternative food networks. Socio.hu Társadalomtudományi Szemle, 145–166
6. Burke MB, Miguel E, Satyanath S, Dykema JA, Lobell DB (2009) Warming increases the risk of civil war in Africa. Proc Natl Acad Sci 106(49):20670–20674

7. Claeys P (2012) The creation of new rights by the food sovereignty movement: the challenge of institutionalizing subversion. Sociology 46(5):844–860
8. Clapp J, Moseley WG (2020) This food crisis is different: covid-19 and the fragility of the neoliberal food security order. J Peasant Stud 47(7):1393–1417
9. Clendenning J, Dressler WH, Richards C (2016) Food justice or food sovereignty? Understanding the rise of urban food movements in the USA. Agric Hum Values 33:165–177
10. Conner DS, Levine R (2007) Circles of association: the connections of community-based food systems. J Hunger Environ Nutrition 1(3):5–25
11. Cone C, Myhre A (2000) Community-supported agriculture: a sustainable alternative to industrial agriculture? Hum Organ 59(2):187–197
12. Corsi A, Barbera F, Dansero E, Peano C (2018) Alternative food networks. McMillan, London
13. Corvellec H, Stowell AF, Johansson N (2022) Critiques of the circular economy. J Ind Ecol 26(2):421–432
14. de Camargo Macedo A, Souza-Esquerdo VF, Borsatto RS (2023) Neoliberal agenda and the dismantling of socially-efficient public food procurement programs: an emblematic case. Glob Food Sec 37:100683
15. Dixon J (1999) A cultural economy model for studying food systems. Agric Hum Values 16:151–160
16. Diekmann LO, Gray LC, Baker GA (2020) Growing 'good food': urban gardens, culturally acceptable produce and food security. Renewable Agric Food Syst 35(2):169–181
17. Facchini E, Iacovidou E, Gronow J, Voulvoulis N (2018) Food flows in the United Kingdom: the potential of surplus food redistribution to reduce waste. J Air Waste Manag Assoc 68(9):887–899
18. Feenstra GW (1997) Local food systems and sustainable communities. Am J Altern Agric 12(1):28–36
19. Friesen LD (2017) The failures of neoliberal food security and the food sovereignty alternative. USURJ: Univ Saskatchewan Undergraduate Res J 4(1)
20. Geng Y, Sarkis J, Bleischwitz R (2019) How to globalize the circular economy. Nature 565(7738):153–155
21. Gleick PH, Lowi MR (1992) Water and conflict. American Academy of Arts and Sciences
22. Goodman D, DuPuis EM, Goodman MK (2012) Alternative food networks: knowledge, practice, and politics. Routledge
23. Güven G, Aysel İ (2016) Food footprint in daily life: Opinions about the consumption of convenience food/Gündelik hayatta gida ayak izi: Hazır gıda tüketimine yönelik görüşler. Eğitimde Kuram ve Uygulama 12(2):403–426
24. Heynen N, Kurtz HE, Trauger A (2012) Food justice, hunger and the city. Geogr Compass 6(5):304–311
25. Himanen SJ, Rikkonen P, Kahiluoto H (2016) Codesigning a resilient food system. Ecol Soc 21(4)
26. Hodgson K (2012) Planning for food access and community-based food systems. American Planning Association, US
27. Hupper A (2017) The role of cooperation and prosocial behavior in food buying clubs: an exploratory study. Honors College 308
28. International Food Policy Research Institute (IFPRI) (2024) Food systems. https://www.ifpri.org/topic/food-systems/#:~:text=Food%20systems%20are%20the%20sum,foods%20to%20consumption%20and%20disposal
29. Jarosz L (2008) The city in the country: growing alternative food networks in Metropolitan areas. J Rural Stud 24(3):231–244
30. Joshi A, Azuma AM, Feenstra G (2008) Do farm-to-school programs make a difference? Findings and future research needs. J Hunger Environ Nutr 3(2–3):229–246
31. Kallis G, Kostakis V, Lange S, Muraca B, Paulson S, Schmelzer M (2018) Research on degrowth. Annu Rev Environ Resour 43:291–316
32. Kibler KM, Reinhart D, Hawkins C, Motlagh AM, Wright J (2018) Food waste and the food-energy-water nexus: a review of food waste management alternatives. Waste Manage 74:52–62

33. Korhonen J, Nuur C, Feldmann A, Birkie SE (2018) Circular economy as an essentially contested concept. J Clean Prod 175:544–552
34. Lee WJY (2023) The political economy of Australia's waste crisis: From neoliberalism to the circular economy agenda. Circular Econ Sustain 3(3):1703–1721
35. Lombardi GV, Atzori R, Acciaioli A, Giannetti B, Parrini S, Liu G (2019) Agricultural landscape modification and land food footprint from 1970 to 2010: a case study of Sardinia, Italy. J Clean Prod 239:118097
36. Lloro T, González F (2022) Food activism and negotiating the gendered dynamics of public cultures of care. Canadian Food Studies/La Revue canadienne des études sur l'alimentation 9(2):180–204
37. Mares TM, Alkon AH (2011) Mapping the food movement: addressing inequality and neoliberalism. Environ Soc 2(1):68–86
38. Marsden T, Morley A (2014) Sustainable food systems. Building a new paradigm. Earthscan from Routledge
39. Martínez-Alier J (2020) A global environmental justice movement: mapping ecological distribution conflicts. Disjuntiva. Crítica de les Ciències Socials 1(2):81–126
40. Martínez-Alier J (2023) The EJAtlas and the making of the world movement for environmental justice. A "sociology of absences"? Tocqueville Rev 44(1):25–48
41. Mees C, Stone E (2012) Food, homes and gardens: public community gardens potential for contributing to a more sustainable city. In: Sustainable food planning: evolving theory and practice. Wageningen Academic, pp 431–452
42. Moreno-Peñaranda R, Kallis G (2010) A coevolutionary understanding of agroenvironmental change: a case-study of a rural community in Brazil. Ecol Econ 69(4):770–778
43. Navas G, D'Alisa G, Martínez-Alier J (2022) The role of working-class communities and the slow violence of toxic pollution in environmental health conflicts: a global perspective. Glob Environ Chang 73:102474
44. Nicolopoulou-Stamati P, Xiangli N, Matiatos I, Kotampasi C, Stamatis PS, Sasco AJ, Hens L (2013) Green economy: the contribution of corporate social responsibility. Mech Econ Regul 2:30–44
45. Niskanen J, Anshelm J, McLaren D (2020) Local conflicts and national consensus: the strange case of circular economy in Sweden. J Clean Prod 261:121117
46. Oba B, Ozsoy Z (2020) Unifying nature of food: consumer-initiated cooperatives in Istanbul. Soc Bus Rev 15(4):349–372
47. Olesen JE (2009) Climate-friendly food. Momentum 4:4–8
48. Ostrom MR (2007) Community supported agriculture as an agent of change. In: Hinrichs C, Lyson T (eds) Remaking the North American food system. University of Nebraska Press, pp 99–120
49. Patel R (2009) Food sovereignty. J Pleasant Stud 36(3):663–706
50. Parnham JC, Laverty AA, Majeed A, Vamos EP (2020) Half of children entitled to free school meals did not have access to the scheme during Covid-19 lockdown in the UK. Public Health 187:161–164
51. Popescu CRG (2021) Measuring progress towards the sustainable development goals: creativity, intellectual capital, and innovation. In: Popescu C (ed) Handbook of research on novel practices and current successes in achieving the sustainable development goals. IGI Global, pp 125–136
52. Schipanski ME, MacDonald GK, Rosenzweig S, Chappell MJ, Bennett EM, Kerr RB, Blesh J, Crews T, Drinkwater L, Lundgren JG, Schnarr C (2016) Realizing resilient food systems. Bioscience 66(7):600–610
53. Stahel WR (2016) The circular economy. Nature 531(7595):435–438
54. Singh R, Khanna V (2021) Climate change and food systems: implications on food security. In: Climate change and resilient food systems: issues, challenges, and way forward, pp 73–111
55. Swensson LF, Tartanac F (2020) Public food procurement for sustainable diets and food systems: the role of the regulatory framework. Glob Food Sec 25:100366

56. Teigiserova DA, Hamelin L, Thomsen M (2020) Towards transparent valorization of food surplus, waste and loss: clarifying definitions, food waste hierarchy, and role in the circular economy. Sci Total Environ 706:136033
57. Thakali A, MacRae JD (2021) A review of chemical and microbial contamination in food: what are the threats to a circular food system? Environ Res 194:110635
58. Turner B, Hope C (2014) Ecological connections: reimagining the role of farmers' markets. Rural Soc 23(2):175–187
59. World Research Institute (2024) Our challenge. https://www.wri.org/food
60. Yıldırım S, Bostancı SH, Yıldırım DC (2024) Examining the resilience of local food systems against food insecurity in sudden crises. In: Food security in a developing world: status, challenges, and opportunities. Springer Nature Switzerland, Cham, pp 355–369
61. Zink T, Geyer R (2017) Circular economy rebound. J Ind Ecol 21(3):593–602
62. Zucchella A, Previtali P (2019) Circular business models for sustainable development: a "waste is food" restorative ecosystem. Bus Strateg Environ 28(2):274–285

One Health in the Americas: Mapping Health and Governance Strategies in the Context of the Joint Action Plan 2022–2026

Rosana Márcia da Costa Silva Herculano, Denilson Teixeira, and André Cavalcante da Silva Batalhão

Abstract The One Health approach, proposed by the United Nations (UN) through its agencies: Food and Agriculture (FAO), Animal Health (WOAH), Health (WHO), and Environment (UNEP), is a concept that interrelates human, animal, and environmental health. Based on this concept, the perspective of this work was to identify health guidelines and strategies grounded in the document One Health Joint Plan of Action (OH JPA) 2022–2026, understood, in this analysis, as the structural basis for the debate and implementation actions. In this context, this study aimed to map and identify the response formulations carried out in the member countries of the Pan American Health Organization (PAHO), one of the WHO's regions of operation, focusing on the analysis of political aspects and legislations, as outlined in strategy 1 of the "Theory of Change" of One Health and presented in the OH JPA (2022–2026). The methodology involved conducting a narrative literature review, aimed at researching and providing an epistemological, conceptual, and temporal analysis of One Health and its articulation and implementation strategies. The Scopus and Web of Science databases were consulted, using descriptors: One Health, Theory of Change, action plan, and governance. Documentary data from the studied countries were researched on the websites of WHO, FAO, UNEP, WOAH, and PAHO. The analysis focused on the six lines of action of the OH JPA 2022–2026. The results showed the participation of countries in monitoring and control systems such as IHR, Codex Alimentarius, and GF-TADs. There are political strategies, legislations, guidelines, and programmes instituted by PAHO member countries for AMR, with a lesser degree of adherence and greater inclusion in food security strategies, animal health, and environmental conservation agreements that configure instruments for promoting the One Health.

Keywords One Health · Theory of change · Action plan · Governance

---

R. M. da Costa Silva Herculano (✉) · D. Teixeira · A. C. da Silva Batalhão
Graduate Programme in Environmental Sciences, Federal University of Goiás, Goiânia, Goiás, Brazil
e-mail: rosana.marcia@discente.ufg.br

A. Cavalcante da Silva Batalhão and P. Singh (eds.), *Blending Approaches Towards Sustainability Focus*, Earth and Environmental Sciences Library,
https://doi.org/10.1007/978-3-031-93518-3_6

1 Introduction

In contrast to the marked processes of techno-scientific and economic development, concentrated in certain regions of the planet, global health is threatened by the emergence and return of infectious diseases. Although advances in reducing and overcoming pathologies can be noted, some long-known diseases continue to impact the rise in mortality rates, especially in regions and countries with lower levels of economic development [23]. It is in this context of a global call to action that the One Health approach emerges, as a joint action initiative of the quadripartite organisation composed of international agencies: the "Food and Agriculture Organization of the United Nations (FAO), the United Nations Environment Programme (UNEP), the World Health Organization (WHO), and the World Organisation for Animal Health (WOAH)" [18], for holistic, integrated, and intersectoral strategies related to human, animal, and environmental health [4].

The scenario of intense and complex challenges, especially socio-environmental ones, where population growth, unplanned urbanisation, continuous deforestation processes, the occupation of native vegetation areas by anthropic activities, combined with the elimination of borders through the globalisation of transport systems and the economy [13], leads to the emergence and spread of diseases with significant impacts on health systems. In this sense, the One Health approach advocates and proposes that strategies for political decisions on control, mitigation, and improvements in monitoring and evaluation processes will only be possible if the interconnection between its dimensions is understood [2].

Although the concept of One Health begins with historically constructed definitions and knowledge, in the context of veterinary medicine, such as the concept of "zoonosis", originating in the late nineteenth century, to define the connections that occur between humans and animals that cause the transmission of infectious diseases [16], the approach was only structured at the beginning of the twenty-first century [1]. For a more precise temporal definition, it can be stated that the One Health approach has been consolidating itself as a planning instrument for disease prevention since 2004, when the 12 Manhattan principles were formulated in New York, no Symposium organised by the "Wildlife Conservation Society at Rockefeller University" [16]. Although the One Health concept has emerged in the veterinary community, it has been discussed and evolved as a scientific foundation, over the last few years, in the most diverse areas of knowledge [15]. Discussions around One Health have expanded due to the emergence of new health emergencies, such as the COVID-19 pandemic and various zoonoses. These include Severe Acute Respiratory Syndrome (SARS), Nipah virus disease, different strains of zoonotic influenza (such as H5N1, H7N9, and the 2009 H1N1 pandemic), arboviral diseases like Zika virus disease, yellow fever, and chikungunya, as well as Ebola virus disease, plague, and Middle East Respiratory Syndrome (MERS-CoV), which threaten the interface between humans, animals, and the environment [5].

If well understood, the One Health approach can be adopted as a political strategy, guiding the understanding that health strategies should be planned and implemented under the integral understanding of the origins of pathologies. They can offer strategic and systematic governance frameworks taking into account the contexts, needs, and priorities of each country [13]. The approach can contribute, beyond health promotion actions, to sustainability promotion strategies focused on the Sustainable Development Goals (SDGs), outlined in the 2030 Agenda, which encompasses environmental, social, economic, and institutional dimensions, for which 17 goals, with targets and indicators for Sustainable Development, are to be implemented at global, regional, national, and local levels [12]. In this sense, the quadripartite organisation offers technical support and proposes that the approach be institutionalised and implemented with solid governance mechanisms that promote multisectoral work to achieve positive results through collaborative actions [13].

It is important, however, that governance strategies are planned so that the One Health approach does not have an unbalanced representation or focus on its dimensions, for example, centred on humans [2], so as not to lose sight of the interrelationship of the dimensions. It is crucial that the environment is seen as a fundamental dimension of the approach and not just as ecosystem services with economic value, as when negatively impacted by anthropic actions, it significantly contributes to the emergence of diseases [4]. It is precisely the holistic nature, capable of encompassing different areas of knowledge, based on scientific and historical roots, that will enable the protection and promotion of global health. Animal, human and environmental health must be viewed in an integrated manner, without hierarchies between human populations, animals and the environment [15] if the idea is to promote a sustainable and lasting ecosystem [2] with the capacity to sustain future generations.

Given the theoretical, conceptual, and methodological characteristics of the One Health approach, it can be consensually considered that it facilitates communication, collaboration, and coordinated work among stakeholders towards promoting the health of people, animals, and ecosystems [1, 2]. It is within this proposed interconnection that international organisations operate, collecting and monitoring data, developing guidance tools, providing technical support, and funding health strategies, indicating as a counterpart the implementation of political and strategic actions of One Health. It is true, however, that there are numerous challenges to implementing the One Health Approach, although it is a proposal with many benefits [15], highlighting the need for actions in education, research, diagnosis, surveillance, and financing.

As an implementation tool of the approach, the One Health High-Level Expert Panel (OHHLEP), created in 2021 under the influence of emergencies caused by the COVID-19 pandemic, with the aim of scientific analyses of health crises and developing long-term strategies [16], developed and approved in 2022 the One Health Joint Plan of Action (OH JPA) 2022–2026, a document that established guidelines and described actions, goals and strategies for the collective implementation of One Health, led by the four UN agencies [5]. The document is proposed as a guiding framework for the approach, as it systematises existing actions, proposes a set of new strategies, and provides political and legislative guidance so that from the perspective of the Theory of Change, collaboration, communication, training processes, and coordination of the implementation of One Health are promoted [5].

The Theory of Change structures the methodological framework of the One Health approach when it includes three pathways for strategic actions, which are: (1) Policy, legislation, advocacy, and funding; (2) Organisational development, implementation, and sectoral integration; and (3) Data, evidence, education, and knowledge exchange [5] indicate sectors to be considered for developing strategic actions that can enhance the capacity to address challenges related to the triad.

Given the structured approach on a global scale, under the supervision of the quadripartite organisation, guided by the already approved legal instruments, published scientific documents, and data collection and monitoring tools made available, the question that emerges is: How are nation-states formulating political responses with a legal basis for the One Health approach, in light of the actions proposed by the quadripartite organisation? This question guided the investigation developed in this study which, in light of this stimulus, aimed to map and identify the response formulations instituted by the member countries of the Pan American Health Organization (PAHO), under WHO guidelines. The mapping was carried out focusing on the analysis of political aspects and legislations, in line with the outline of strategy 1 of the "Theory of Change" of One Health and presented in the OH JPA—2022–2026. The definition of the spatial cut is based on the importance of the region due to its territorial, demographic, economic, and environmental dimensions in relation to the planet.

There are numerous and varied initiatives and work fronts in view of the One Health approach, developed by international organisations, to mobilise the understanding and practice of strategies and actions to be implemented by countries, hence, observing the response formulations helps to identify advances and weaknesses in the context. In this sense, the study sought to identify and understand how the approach is being operationalised in practice, identifying responses already formulated for One Health through the national legislation in force in PAHO countries, having as a theoretical basis the OH JPA, which indicates the implementation of regulatory frameworks in member countries as the first concrete action strategy for the realisation of One Health.

2 Methods

The methodological proposal of this study was developed in three stages, as presented in the study design flowchart—Fig. 1.

The first stage consisted of a narrative literature review, with the objective of researching and understanding the conceptual and temporal frameworks of the One Health approach and its articulation and implementation strategies. The narrative review adopts a broader analysis methodology [17], and thus enabled the identification of peer-reviewed articles that contributed to the understanding of the concept and implementation strategies, especially those related to the validation of legislation that legitimises the approach.

From the perspective of health control, the WHO provides, based on the International Health Regulations (IHR), the e-SPAR/WHO Platform, available at: https://extranet.who.int/e-spar, which aims to collect and compare data to evidence response capacity regarding health safety. The tool is based on the commitment of Member States to present annual self-assessment reports on IHR implementation, from which they receive support for meeting their obligations.

An important instrument in combating antimicrobial resistance is the Global Action Plan on AMR (GAP—AMR), approved in 2015, by WHO Member States [21], which has since guided political and governance actions at national levels.

The plan's broad objective is to ensure the responsible and safe use of antimicrobials. Countries need to develop and approve their national AMR plans to align with the "Global Action Plan" [6], and this can improve capacity for action, expand understanding and awareness of AMR, strengthen surveillance systems, reduce infections; increase control over antimicrobial use and optimise investments in diagnostics, medicines, and vaccination [3]. Global and national planning strategies are important as they aim to prevent deaths and avoid regression to the pre-antibiotic era [11].

The concrete execution of the plan's actions includes surveillance tools. Monitoring of antimicrobial consumption (AMC) and resistance (AMR) on a global scale is carried out through the Global Antimicrobial Resistance and Use Surveillance System (GLASS) to support efforts at the global level and the development of national surveillance systems [22]. For monitoring antimicrobial resistance in the Americas, the Latin American AMR Surveillance Network (ReLAVRA) was established in 1996 through PAHO's Special RAM Programme. The system aims to improve and disseminate knowledge and increase awareness about AMR in the region.

In addition to monitoring systems, guidance documents for RAM data collection are provided to countries [7], in this case, for both human and animal health. For data related to animal health, the OIE Observatory is available, a system for collecting, observing, and analyzing data from different information sources based on each country's self-assessment, which allows tracking progress and identifying challenges [7].

Following the characterisation, it was observed that environmental aspects are being addressed through debate and the establishment of international agreements. The environmental debate related to One Health reiterates the existence of environmental issues, such as rising global temperatures, deforestation, and carbon dioxide emissions, which significantly contribute to the emergence of diseases. Although the debate is old, the first decades of the twenty-first century mark the integration of this dimension into the One Health context, with strategic action milestones including major global environmental conferences and the signing of important treaties such as the 2015 Paris Agreement.

3.3 Formulating Responses for One Health

After identifying the mechanisms for implementing One Health by the quadripartite organisation, the next step was to identify how and which political and governance responses, at national levels, are being executed by PAHO countries. Observing, at first, the national responses presented to the programmes and systems instituted and already consolidated by the quadripartite organisation, it was noted that there is a connection between the action strategies being implemented by PAHO countries and the One Health dissemination programmes of international agencies, which constitute political responses. The data shows, as presented in Table 1, significant involvement in the development of actions for combating transboundary diseases, health control, with high levels of participation from the countries. Actions aimed at combating antimicrobial resistance demonstrate the countries' participation and responses are still incipient, despite already being instituted programmes since 2015.

The data reveal that out of the 35 countries analysed, only 15 participate in the Regional Steering Committee of the Global Framework for the Progressive Control of Transboundary Animal Diseases (GF-TADs) [14], which is a joint action by FAO and OIE, with WHO participation, aiming to promote alliances for the prevention and control of the spread of transboundary diseases.

Regarding the International Health Regulations (IHR), the data identified on the e-SPAR/WHO Platform show that all PAHO member countries submitted annual self-assessment reports on IHR implementation for the year 2023. It is noted that compared to the regional and global averages of the Americas, which show a response capacity of 64% for 2023, 18 countries are at or above this result, and 17 have lower results. It stands out that countries with higher percentages of health control capacity are those with better socioeconomic indicators.

The data on the control and combat of antimicrobial resistance show that in terms of monitoring antimicrobial consumption (AMC), only Peru and Colombia develop data control strategies. Antimicrobial resistance (AMR) control actions are carried out by Argentina, Brazil, Canada, Colombia, Paraguay, Trinidad and Tobago, and the United States. Despite the fragility in the effectiveness of strategies to control antibiotic consumption and antimicrobial resistance, the number of countries integrated into ReLAVRA, the regional system of the Americas created to improve the dissemination of knowledge and awareness about AMR, is significant, with 20 countries associated with this initiative.

With the purpose of expanding the identification of response formulations, the institutionalised legal bases in PAHO member states were analysed. Through the database provided by FAO, which presents the existing legislation in each country of the analysed region. Given the lack of thematic standardisation of the legal base, since there is a diversity of legal documents instituted in the countries, for various themes, under the most diverse nomenclatures, the search was mainly aimed at identifying National Plans for implementing strategies related to the action lines of the One Health Global Action Plan: One Health, Animal Health, Antimicrobial Resistance,

Table 1 Participation of countries in One Health programs and control systems

Country	Regional Steering Committee of the Global Framework for the Progressive Control of Transboundary Animal Diseases (GF-TADs)	Capacity to respond to the International Health Regulations—(IHR)—2023 (%)	Subscribed GLASS (AMC)	Subscribed GLASS (AMR)	ReLAVRA participation—AMR
Antígua and Barbuda		56			
Argentina	✓	64		2019	✓
Bahamas		69			
Barbados	✓	69			
Belize		54			✓
Bolivia		48			✓
Brazil	✓	75		2017	✓
Canada	✓	96		2016	
Chile	✓	83			✓
Colombia	✓	66	2021	2021	✓
Costa Rica	✓	65			✓
Cuba	✓	96			✓
Dominica		61			
Dominican Republic	✓	52			✓
Ecuador	✓	61			✓
El Salvador		92			✓
Grenada		52			
Guatemala	✓	32			✓
Guyana		31			

(continued)

Table 1 (continued)

Country	Regional Steering Committee of the Global Framework for the Progressive Control of Transboundary Animal Diseases (GF-TADs)	Capacity to respond to the International Health Regulations—(IHR)—2023 (%)	Subscribed GLASS (AMC)	Subscribed GLASS (AMR)	ReLAVRA participation—AMR
Haiti		54			
Honduras		58			✓
Jamaica	✓	78			
Mexico	✓	79			✓
Nicaragua	✓	60			✓
Panama	✓	67			✓
Paraguay		62			✓
Peru		36	2019	2019	✓
Saint Kitts and Nevis		69			
Saint Lucia		67			
Saint Vincent and the Grenadines		59			
Suriname		47			
Trinidad and Tobago	✓	65		2020	
United States of America	✓	91		2016	
Uruguay	✓	76			✓
Venezuela		46			✓

Source Pan American Health Organization (PAHO) and World Health Organization (WHO); Food and Agriculture Organization of the United Nations (FAO) and World Organization for Animal Health (WOAH) [14]; United Nations Environment Programme (UNEP) [19]

Food Safety, Biodiversity Conservation, and Climate Change Mitigation. The results obtained are presented in Table 2.

The results made it possible to identify the still incipient existence of National Plans for antimicrobial resistance and a much more structured approach for food safety. Adherence to data monitoring systems on food safety is almost total. Specific legislation or national plans for animal health and biodiversity conservation, and multilateral agreements for climate control were identified.

It is observed that, in general, countries are in the process of structuring the legal basis for the implementation of One Health strategies. Only Cuba, Haiti, and Saint Vincent and the Grenadines do not have a National Action Plan for Animal Health. All countries have National Action Plans for food safety and integrate, even with adherence at different times, the Codex Alimentarius. The implementation of National Plans for antimicrobial resistance control is still fragile. The analysed countries have legislation for biodiversity conservation and have ratified the Paris Agreement, although in different years, through Nationally Determined Contributions (NDC) for the control, reduction, and mitigation of greenhouse gas emissions. It is evident that from a legal standpoint, the One Health dimensions are covered in most of the countries analysed.

4 Discussion

Despite the discussion and debate context about the One Health approach being amplified over the first two decades of the twenty-first century, the occurrence of the Covid-19 pandemic drew attention to the need to face global health threats from a systemic perspective, including instituting structures for integrating evidence, currently fragmented, so that the interconnection between human, animal, and environmental aspects is understood and facilitates the determination of priorities and decision-making [23]. The pandemic, due to its proportions and consequences, promoted multiple changes but also interfered with processes that were already underway, such as the RAM agenda, for example, causing stagnation [1].

In light of the numerous proposed actions carried out by the quadripartite organisation's agencies, it is generally observed that there are responses being implemented by national governments, but with a great need for increased technical support and planning of national coordination for prioritising One Health agendas. This condition is described in the OH JPA, which states that many obstacles prevent the practical implementation of One Health, as it is necessary to address technical, institutional and professional barriers, as well as competing priorities and funding deficits [5].

Overcoming the challenges will require governance efforts and the structuring of political strategies to have positive influences on decision-making. The context requires the dissemination of knowledge and understanding about the One Health approach to expand the vision of the future and the threats currently underway. The One Health perspective must be appropriated from a scientific viewpoint, with

Table 2 Countries' responses to the One Health approach, through legislation

Country	Animal health and welfare	National Action Plan against AMR	Food Safety Policy	Codex Alimentarius (FAO/WHO)—Year of accession	Action plan for biodiversity conservation	iNDC—Nationally Determined Contribution for the period 2020–2030
Antígua and Barbuda	✓		✓	1988	✓	2018
Argentina	✓	✓	✓	1963	✓	2018
Bahamas	✓		✓	2002	✓	2015
Barbados	✓	✓	✓	1970	✓	2015
Belize	✓		✓	1992	✓	2020
Bolivia	✓		✓	1971	✓	2018
Brazil	✓	✓	✓	1968	✓	2018
Canada	✓	✓	✓	1963	✓	2021
Chile	✓		✓	1969	✓	2022
Colombia	✓	✓	✓	1969	✓	2019
Costa Rica		✓	✓	1970	✓	2020
Cuba			✓	1964	✓	2020
Dominica	✓		✓	1990	✓	2020
Dominican Republic	✓		✓	1971	✓	2018
Ecuador	✓	✓	✓	1970	✓	2018
El Salvador	✓	✓	✓	1975	✓	2021
Grenada	✓		✓	1982	✓	2019
Guatemala	✓		✓	1968	✓	2018
Guyana	✓		✓	1970	✓	2018
Haiti			✓	1984	✓	2018
Honduras	✓		✓	1988	✓	2015
Jamaica	✓			1971	✓	2019
Mexico	✓		✓	1969	✓	2018
Nicaragua	✓		✓	1971	✓	2019
Panama	✓		✓	1972	✓	2018
Paraguay	✓	✓	✓	1969	✓	2018
Peru	✓	✓	✓	1963	✓	2018
Saint Kitts and Nevis	✓		✓	1996	✓	2019
Saint Lucia	✓		✓	1987	✓	2018

(continued)

Table 2 (continued)

Country	Animal health and welfare	National Action Plan against AMR	Food Safety Policy	Codex Alimentarius (FAO/WHO)—Year of accession	Action plan for biodiversity conservation	iNDC—Nationally Determined Contribution for the period 2020–2030
Saint Vincent and the Grenadines			✓	2004	✓	2019
Suriname	✓		✓	1984	✓	2015
Trinidad and Tobago	✓		✓	1964	✓	2018
United States of America	✓	✓	✓	1963	✓	2021
Uruguay	✓	✓	✓	1970	✓	2018
Venezuela	✓		✓	1969	✓	2015

Source Pan American Health Organization (PAHO); Food and Agriculture Organization of the United Nations (FAO); United Nations Environment Programme (UNEP) [19]

intense political will and active civil society participation, as recommended by the Rome Declaration [15].

The interdisciplinary aspect must be a driving force for actions related to One Health, given the nature of the approach, as it can contribute to overcoming challenges, especially regarding the difficulties of communication frequency and quality, demonstrating the idea of multiple benefits. Knowledge production is also a determining factor, and in this case, it would be fundamental for the One Health approach to integrate academic curricula to expand debates, discussions, and strategic actions and especially the multidisciplinary culture [18].

The Global Action Plan is an important document, capable of contributing to consolidating the implementation of the One Health approach at regional, national, and even local levels. It is interesting to note that it systematises existing strategies and adds new proposals that cover human, animal, and environmental health areas, as it was developed after discussions and many One Health strategies were already underway. Although the Global Action Plan was instituted in 2022, and a guiding guide was presented in 2023 for countries to have a basis for planning their national One Health plans, the actions are still very preliminary. No implemented national action plans were identified. Countries are still structuring work teams for proposal development.

Given the context, it is highlighted that the responses do not constitute effective results of the Global Action Plan implementation. Many actions were even initiated before the plan's implementation, but there is no doubt about the instrument's ability to expand understanding and execution of strategic processes to consolidate the One Health approach.

Looking at what is already being executed is important to understand the advances and, above all, gaps in managing public health risks and proposing improvements for a sustainable future [2]. The participation of countries in data collection programmes and systems is significant, but there are high levels of response fragility, especially regarding combating antimicrobial resistance.

It is possible to observe that, under the recommendation of PAHO and WHO guidelines, actions and strategies aimed at implementing One Health, especially in relation to health security, food safety and antimicrobial resistance. For the implementation of the strategies, examples include the adherence of countries to the "International Health Regulations (IHR), Codex Alimentarius, adherence to international health safety standards of the OIE and actions convergent with the Global Action Plan on Antimicrobial Resistance" [13]. The strategies described have, respectively, as main objectives: "Defining responsibilities for managing public health events and emergencies with cross-border potential; defining standards and guidelines and other internationally recognized recommendations on food, food production and food safety; defining health safety standards in the international trade of terrestrial and aquatic animals; monitoring and evaluating decisions, actions and strategies to combat antimicrobial resistance" [13].

The analysis of the data revealed the existence of an organization involving countries with the goal of implementing a global strategy for the progressive control of transboundary animal diseases, which facilitates actions to mitigate African swine fever, avian influenza, and rinderpest. For controlling the spread of transboundary diseases, the results point to the need for expanding technical support strategies to involve countries not yet included in this system, as non-involvement impacts human, animal, and environmental health structures, with significant socio-environmental repercussions.

Voluntary participation in Codex Alimentarius standards, guidelines, and food codes signals the construction of international food safety mechanisms. However, it is important for countries to legally regulate these international standards in their territories, thus creating standardised planning and execution rules for food trade definitions and requirements, which is crucial for all themes related to One Health.

Although antimicrobial control has been part of actions implemented since 2015, it is still very fragile. There is a need for political actions to integrate countries that have not yet been able to implement their national strategies. The increase in RAM is a challenge that needs to be prioritised as it relates to environmental, human, and animal transformation processes that can, through resistance caused by intense exposure to antimicrobials, lead to a regression to the pre-antibiotic era [11].

Facing RAM challenges is closely related to the priorities established by political actions. A good example is Brazil, which joined the WHO's Global Antimicrobial Resistance and Surveillance System (GLASS) in 2017 and developed BR-GLASS in 2018, approved and published its sectoral plan and instituted the National Program for the Prevention and Control of Antimicrobial Resistance in Agriculture, influenced by the National Action Plan published in 2019, but with changes in government representations, the working committees were dissolved, leading to the stagnation of the agendas already instituted [1].

The facts show the close correlation between the AMR agenda and the One Health approach. The One Health approach presents the best strategies to combat the current AMR crisis, with an emphasis on interdisciplinary and cooperative actions [10]. To complement the global agenda, an important step is the structuring, at national levels, of action plans to combat AMR, an action that is already part of an agreement reached in 2015 during the "WHO World Health Assembly, the FAO Conference, and the OIE World Delegate Assembly "[7].

The action is crucial as, globally, antimicrobial resistance (RAM) surveillance acts as an early warning system, providing emerging data essential for identifying long-term trends. At the national level, this surveillance guides policymakers, helping them implement appropriate and timely interventions. Locally, it empowers field professionals to make coherent decisions on treating infectious diseases in animals and plants. To achieve these objectives, the implementation of an efficient system for collecting and analyzing data on pathogens [10] can be an excellent support tool.

RAM control strategies must consider the need to establish policies and strategic planning, create management systems and prescription of medicines, technology for optimised prescription and use of antimicrobials, cultural and behavioural transformation, and political and civil society involvement [3].

Regarding existing legislation, although legal documents composed of National Plans for some themes and legislation for strategy implementation have been identified, it cannot be affirmed that they are linked as effective responses to One Health approach guidelines prescriptions. Much needs to be done in structuring a standardised legal basis to improve One Health understanding and communication and facilitate planning and strategic action development. Compatibility guided by international standards consolidates the common approach and creates references for consistent decision-making, aiming for progress and achieving objectives [7].

Environmental data on RAM are difficult to translate into policies; there is a growing process of knowledge dissemination about antimicrobial environmental exposures, but systematic studies on medium- and long-term threats are lacking, compromising the understanding of human and animal health impacts [11].

Challenges, however, should not be an impediment to planning knowledge promotion actions and strategy development. From this perspective, the OH JPA can be considered the basis for implementing and operationalizing One Health. The proposed actions should be an initial milestone for expanding the strategies for solid implementation of One Health [5]. The imminent risks of emerging unknown diseases that could threaten human, animal, and environmental health and well-being cannot be disregarded.

5 Conclusion and Relevant Reflections

The development of this study evidenced that the consolidation of the One Health approach relies on the existence of a consolidated global structure capable of providing reliable technical support for guiding national strategies. However, there are weaknesses in prioritising the understanding and structuring of strategic actions of the approach.

The study provided important information capable of contributing to the purpose of identifying global proposals and national responses, as knowing to what extent countries implement international standards and how they do it is an important step [7] for identifying progress, gaps, and weaknesses. The implementation of the One Health approach has proven challenging, but understanding the conditions of PAHO member countries, which coexist with the risks of zoonoses emergence, mainly associated with climate change, deforestation, and large-scale urbanisation, the conditions of interaction between government agencies, lack of resources, and communication [20], is a decisive step for understanding the need to establish strategic action plans for the approach.

It is worth highlighting that maintaining global health justifies the need to develop training actions in preparation for and response to sustainable emergencies, a condition that requires investment in infrastructure and human capital so that preparation for emergency response is continuous [23]. In this sense, strategies must be executed locally to compose a sustainable or at least resilient global scenario.

An important conclusion of this study is that there are responses. Countries have implemented legal bases. Some are more structured, others more fragile as identified for RAM, and are integrated into data collection and monitoring systems for strategic actions for One Health dimensions. However, the existing responses need to be integrated into more robust ones, especially related to the consolidation of coordination, planning, and cooperation structures.

Certainly, the work presents limitations, and in this case, the analysis of a limited number of situations of countries' participation in One Health programmes and monitoring and control systems can be cited, given the diversity of systems and difficulty in identifying through digital means. Another limitation is the restriction of the analysis of legal documents to National Plans, due to the diversity of nomenclatures and the high volume of documents. Certainly, a more detailed look at the legal bases and other actions executed in each country could highlight the existence of other responses to the One Health approach already underway, better defining their stages of action in this context.

The OH JPA envisages as a desired impact a world capable of preventing, predicting, detecting, and responding to health threats and improving the health of humans, animals, plants, and the environment, while contributing to sustainable development [5] and it is towards this direction that One Health governance strategies should converge.

5.1 Theoretical Implications

The findings underscore the importance of adopting a systemic perspective to address global health threats, which aligns with the One Health approach. This perspective emphasises the interconnectedness of human, animal, and environmental health, advocating for integrated strategies to enhance understanding and decision-making processes. The research highlights the need for a comprehensive legal framework and robust governance structures to facilitate the implementation of One Health strategies. These theoretical insights contribute to the existing body of knowledge by validating the significance of a multidisciplinary approach in addressing complex health challenges and promoting sustainable development.

5.2 Practical Implications

This research reveals the necessity for countries to develop and implement structured national plans for One Health, focusing on antimicrobial resistance, food safety, animal health, and biodiversity conservation. The identification of gaps and weaknesses in current strategies underscores the need for continuous technical support and capacity building. The findings suggest that effective communication, collaboration, and coordination among government agencies and international bodies are crucial for the successful implementation of One Health initiatives. These practical implications can guide policymakers and health practitioners in enhancing the efficiency and effectiveness of their health strategies.

5.3 Implications for Global Sustainability

The study highlights the critical role of One Health in promoting global sustainability. By integrating human, animal, and environmental health, the approach contributes to the achievement of the SDG. The research suggests that addressing health threats through a One Health lens can mitigate the impacts of climate change, deforestation, and urbanisation on health systems. The findings advocate for sustainable investments in infrastructure and human capital to ensure resilient health systems capable of responding to emergencies. These implications reinforce the need for a global commitment to sustainable practices that protect and promote the health of all living beings and the environment.

5.4 *Suggesting a One Health Action Plan*

The objective of the One Health Action Plan is to implement a comprehensive strategy that integrates human, animal, and environmental health. To achieve this, the plan involves several key actions. Firstly, it is essential to develop and implement structured national plans focusing on antimicrobial resistance, food safety, animal health, and biodiversity conservation. Strengthening and standardising legal frameworks to support One Health strategies is also crucial, ensuring these frameworks comply with international standards. Continuous technical support and capacity building will be provided to ensure the effective implementation and monitoring of One Health initiatives. Furthermore, improving communication, collaboration, and coordination among government agencies and international bodies is vital. This will enhance the efficiency and effectiveness of the initiatives. Sustainable investments in infrastructure and human capital will be promoted to create resilient health systems capable of responding to emergencies.

The plan also calls for regular evaluation and adaptation of strategies based on emerging data and feedback to ensure ongoing improvement and effectiveness. By addressing these elements, the One Health Action Plan aims to foster a holistic approach to global health challenges, ensuring sustainable and resilient outcomes for all living beings and the environment.

5.5 *Future Research Agenda in One Health*

Future research should focus on exploring the implementation and impact of One Health strategies at regional, national, and local levels. Investigating the effectiveness of integrated surveillance systems and the role of legal frameworks in supporting One Health initiatives is essential. Additionally, research should address the socioeconomic factors influencing the adoption of One Health approaches and the barriers to their implementation. There is also a need for longitudinal studies to assess the long-term benefits of One Health strategies on global health and sustainability. This future research agenda will provide valuable insights to further refine and enhance the One Health approach.

References

1. Adeyemi OA, Agbabiaka TO, Sujon H (2024) Global One Health post-graduate programmes: a review. One Health Outlook 6. https://doi.org/10.1186/s42522-024-00097-6
2. Cameron A, Esiovwa R, Connolly J et al (2022) Antimicrobial resistance as a global health threat: the need to learn lessons from the COVID-19 pandemic. Glob Policy 13:179–192. https://doi.org/10.1111/1758-5899.13049
3. Charani E, Mendelson M, Pallett SJC et al (2023) An analysis of existing national action plans for antimicrobial resistance-gaps and opportunities in strategies optimising antibiotic use in

human populations. Lancet Global Health 11:E456–E464. https://doi.org/10.1016/S2214-109X(23)00019-0
4. De Leeuw E, Kickbusch I, Rüegg SR (2024) A health promotion perspective on One Health. Can J Public Health-Revue Canadienne de Sante Publique 115:271–275. https://doi.org/10.17269/s41997-024-00872-y
5. FAO; UNEP; WHO; World Organisation for Animal Health (WOAH) (founded as OIE) (2022) One Health Joint Plan of Action, 2022–2026. FAO; UNEP; WHO; World Organisation for Animal Health (WOAH) (founded as OIE)
6. Ferdinand AS, Coppo MJC, Howden BP, Browning GF (2023) Tackling antimicrobial resistance by integrating One Health and the sustainable development goals. BMC Glob Public Health 1. https://doi.org/10.1186/s44263-023-00003-8
7. Ferreira JP, Gochez D, Jeannin M et al (2022) From OIE standards to responsible and prudent use of antimicrobials: supporting stewardship for the use of antimicrobial agents in animals. JAC Antimicrob Resist 4. https://doi.org/10.1093/jacamr/dlac017
8. Food and Agriculture Organization of the United Nations (FAO), World Organisation for Animal Health (WOAH) (2024) Global Framework for the Progressive Control of Transboundary Animal Diseases (GF-TADs). https://www.gf-tads.org/about/en/. Accessed 9 June 2024
9. Jesudason T (2022) A new One Health Joint Action Plan. Lancet Infect Dis 22:1673. https://doi.org/10.1016/S1473-3099(22)00751-4
10. Keck N, Treilles M, Gordoncillo M, et al (2023) A systematic approach toward progressive improvement of national antimicrobial resistance surveillance systems in food and agriculture sectors. Front Vet Sci 9. https://doi.org/10.3389/fvets.2022.1057040
11. Kirchhelle C (2023) The Antibiocene—towards an eco-social analysis of humanity's antimicrobial footprint. Humanit Soc Sci Commun 10. https://doi.org/10.1057/s41599-023-02127-6
12. Kronemberger DMP (2019) The challenges of constructing global SDG indicators. Ciência e Cultura 71(1):40–45
13. Pan American Health Organization (2021) 59th Directing Council 73rd Session of the Regional Committee of WHO for the Americas. PAHO, Washington, D. C.
14. Pan American Health Organization (PAHO), World Health Organization (WHO) Pan American Health Organization (PAHO). https://www.paho.org/en/. Accessed 9 June 2024
15. Di Paolo T (2022) Considerations on the sidelines of the second principle of the Rome Declaration: the challenge of the One Health concept on the health of the future. Int J Risk Saf Med 33:117–124. https://doi.org/10.3233/JRS-227003
16. Van de Pas R (2023) One Health: what's the problem? Development (Basingstoke) 66:191–198. https://doi.org/10.1057/s41301-023-00398-9
17. Talukder B, VanLoon GW, Hipel KW (2022) Planetary health & COVID-19: a multi-perspective investigation. ONE HEALTH 15. https://doi.org/10.1016/j.onehlt.2022.100416
18. Tidman R, Fahrion AS, Thumbi SM et al (2023) United against rabies forum: the first 2 years. Front Public Health 11. https://doi.org/10.3389/fpubh.2023.1010071
19. United Nations Environment Programme (UNEP) (2024) Office for Latin America and the Caribbean in Brazil. https://www.unep.org/regions/latin-america-and-caribbean. Accessed 28 May 2024
20. Wakimoto MD, Menezes RC, Pereira SA et al (2022) COVID-19 and zoonoses in Brazil: environmental scan of one health preparedness and response. ONE HEALTH 14. https://doi.org/10.1016/j.onehlt.2022.100400
21. World Health Organization (2015) Global action plan on antimicrobial resistance. World Health Organization, Geneva. ISBN 978-92-4-150976-3
22. World Health Organization (2022) GLASS guide for national surveillance systems for monitoring antimicrobial consumption in hospitals. World Health Organization, Geneva. ISBN 978-92-4-004156-1
23. Zhang XX, Jin YZ, Lu YH et al (2023) Infectious disease control: from health security strengthening to health systems improvement at global level. Glob Health Res Policy 8. https://doi.org/10.1186/s41256-023-00319-w

Assessing Solid Waste Generation and Environmental Impacts

Jagriti Patel, Hardik Giri Gosai, Sanskriti Mujumdar, and Pradeep Mankodi

Abstract The issue of waste management becomes more and more pressing due to population growth and urban development expansion. With limited land resources and a constant increase in waste generated, decision-makers are facing a daunting challenge in effectively handling and disposing of this ever-growing amount of waste. The waste management issue is particularly severe in urban areas, where the population density is high and the space is at a premium. This expansion in waste and its improper management is one of the main reasons for the ongoing environmental impacts faced globally. This particular issue must be taken into consideration due to its significant environmental impacts on the local, regional, and global scales. Additionally, it is situated near people, potentially posing health risks. Furthermore, it has potential value in terms of recovery through circular economic supply chains. It has become top priority for many cities and governments to find sustainable solution to this problem. The importance of understanding the amount of waste generated globally is essential to comprehend the extent and intricacies associated with solid waste management (SWM). Policymakers and waste management professionals must develop effective strategies for the collection, transportation, treatment and disposal of waste. This chapter provides an overview of the global and Indian solid waste situations. The impact of waste on people and the environment is then discussed, along with a few case studies of trash-related issues and management initiatives.

Keywords Solid waste management (SWM) · Environmental impact · Remediation techniques · Disposal strategies

J. Patel · H. G. Gosai (✉)
Department of Environmental Studies, Faculty of Science, The Maharaja Sayajirao University of Baroda, Vadodara, Gujarat, India
e-mail: hardikgosai182@gmail.com

S. Mujumdar
Department of Civil Engineering, Faculty of Technology and Engineering, The Maharaja Sayajirao University of Baroda, Vadodara, Gujarat, India

P. Mankodi
Division of Marine and Fresh Water, Department of Zoology, Faculty of Science, The Maharaja Sayajirao University of Baroda, Vadodara, Gujarat, India

A. Cavalcante da Silva Batalhão and P. Singh (eds.), *Blending Approaches Towards Sustainability Focus*, Earth and Environmental Sciences Library,
https://doi.org/10.1007/978-3-031-93518-3_7

1 Introduction

The increasing population of human and urbanization have led towards steady increase in the quantity of waste being generated. Manufacturing processes, industries, and municipal solid wastes (MSW) all produce waste products. Waste from commercial, industrial, and domestic sources includes food scraps, yard clippings, nondurable items, containers, and packaging, as well as various inorganic wastes [22, 31]. Wastewater treatment, sewage treatment, and chemical and biochemical processing are all included in liquid waste management. There are differences in waste management techniques between developed and developing countries, between urban and rural locations, and between residential and industrial producers [21, 93]. One of the fundamental utility services supporting twenty-first century civilization, especially in metropolitan areas, is waste management. It is possible to see waste management as a basic human right in addition to a fundamental human necessity. For society and the economy to function properly, access to food, energy, housing, clean water, transportation, and communications must coexist with adequate sanitation and solid waste management (SWM). Despite this, waste management frequently has a lesser public and political profile than other utility services [95].

The costs to society and the economy as a whole of taking little or no action to address waste management can be enormous [96]. When there are no waste laws in place, or when they are not strictly enforced, a waste generator will often take the smallest expensive option. To reduce the annoyance of collected garbage piles, domestic solid waste can be burnt, and disposed of on the street, on unoccupied land, or into sewers, streams, or other water courses. On a big scale, when noteworthy quantities of municipal or industrial solid waste are disposed or burned in open, the adverse effects on air, surface and groundwater, soil and the coastal and marine environment, and thus indirectly on public health, can be dreadful [13, 18]. Residents' health is directly impacted by not having a solid waste collection service, especially children [70]. Waste burning without control releases persistent organic pollutants and particulates that are extremely harmful both locally and worldwide. Flooding is mostly caused by accumulated garbage and clogged drains, which enable vectors to reproduce and transmit dangerous illnesses like dengue fever and cholera [37].

Uncontrolled dumpsites, particularly especially those where hazardous and other wastes are mixed, can spread illness to garbage workers and nearby communities [90]. Groundwater and surface water can get contaminated by dumpsites. These locations may immediately damage rivers and the sea since they are frequently next to them, as well as the coastal environment. One source of marine litter is erosion from coastal dumpsites. Losses from diminished tourism as a result of contaminated beaches and losses from harm to fisheries are two other possible damage costs. One of the main types of polluted sites is former dumpsites, especially ones that have handled hazardous material [15, 25]. This chapter includes an overview of the solid waste scenario in India and worldwide. A few case studies of waste issues and management programs are also included, after which the effects of waste on people and the environment are discussed.

2 Global Status of the Waste Generation Scenario

The management of solid waste is a global issue that impacts everyone on the planet. The decisions made by governments and individuals regarding waste management and consumption have an adverse impact on the general cleanliness, well-being and productivity of communities [69]. Improperly managed waste contaminates the world's oceans [79], which also clogs drains and causes flooding, breeds vectors that spread disease, increases respiratory issues due to airborne particles from burning waste, harms unintentionally consumed animals, and hinders economic development by, for example, reducing tourism. Immediate action is required at all societal levels to direct the unmanaged and badly managed waste from decades of economic expansion [33, 53].

Up till 2016, the yearly global production of municipal solid garbage was 2.01 billion tonnes. The globe is predicted to produce 2.59 billion tonnes of garbage a year by 2030. By 2050, there will be 3.4 billion tonnes of garbage worldwide, a 70% increase, unless immediate action is taken. Overall, 17% of trash is made up of paper and cardboard, followed by food and greens at 14%, plastic at 12%, metal at 4%, rubber and leather at 2%, glass at 2%, wood at 2%, and other at 2% (Fig. 1). According to Fig. 2, East Asia and the Pacific produce 468 million tonnes of waste annually, which is followed by North America (289 million tonnes), Latin America and the Caribbean (231 million tonnes), Europe and Central Asia (392 million tonnes), South Asia (334 million tonnes), Sub-Saharan Africa (174 million tonnes), and the Middle East and North Africa (129 million tonnes), in that order [48].

Since the environment became a global concern in the 1960s, many industrialized nations have made significant progress in managing trash, and the whole community may benefit from a wealth of well-executed case studies [27]. Inevitably the earlier emphasis was on garbage disposed of by contrast, the present focus is upward, addressing the root of the problem [87]. Examples include designing out waste, prevention of its generation, minimizing and reusing, and, in the event that residuals do arise, keeping them concentrated and separate to preserve their intrinsic value for recycling and recovery and to keep them apart from other waste that still has a financial value for recovery [52, 72]. The modified terms 'waste and resource management' and 'resource management', as part of the 'circular economy', reflect

Fig. 1 Overall percentage outlook of waste composition

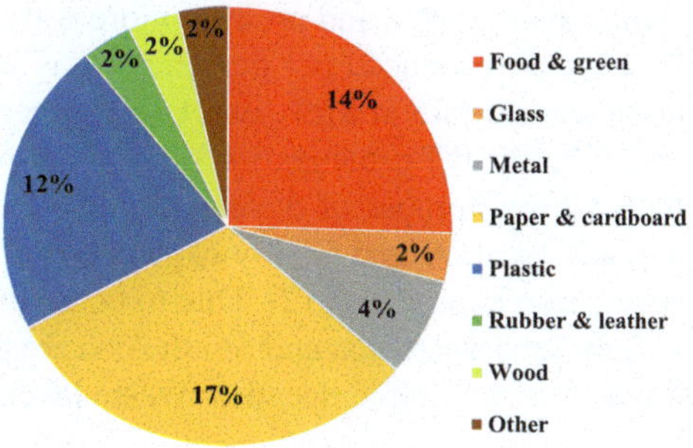

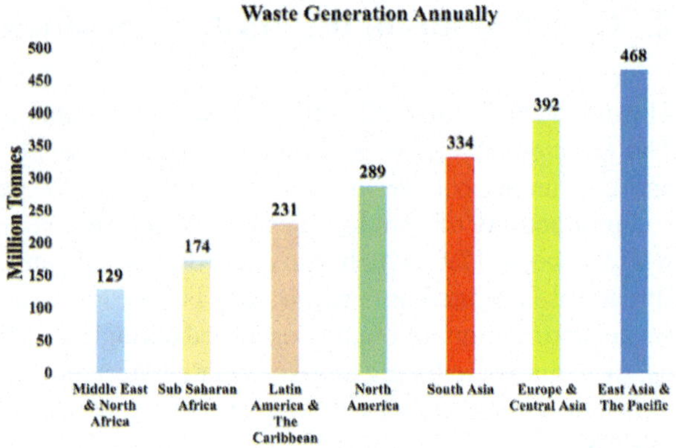

Fig. 2 Global outlook for annual waste generation by region [48]

the objective of changing the underlying mindset from 'waste disposal' to 'waste management' and from 'waste' to 'resources' [71, 88, 91].

Ensuring widespread access to garbage collection services, doing away with uncontrolled dumping and burning, and promoting ecologically sound waste management are still significant difficulties for low- and middle-income nations [29]. Forecasts indicate that over the next 20 years or more, the population of large cities in the lowest-income nations is expected to double, making trash management a more pressing local political concern [36]. This makes achieving this problem much more challenging. To encourage waste avoidance and stop the steadily rising amount of trash produced per person as economies grow, low- and middle-income nations must create and put into action creative and practical policies and practices [50].

With an average daily trash production per person of 2.21 kg, Bermuda, Canada, and the United States are the three nations in the North America area that create the most garbage. All three of these nations have high incomes [48]. The least amount of garbage is produced per person in the three areas with the highest percentage of low- and middle-income countries: Sub-Saharan Africa produces 0.46 kg on average per day, South Asia produces 0.52 kg per day, and East Asia and Pacific produces 0.56 kg per day. In 2016, the projected global average was 0.74 kg of garbage per person per day, and the total amount of solid waste generated was around 2.01 billion tons [6, 48].

Waste creation is accelerated by urbanization. Developed nations and economies are more urbanized and produce more waste overall and per person [12]. Regionally, 2.21 kg are produced per person per day in North America, which has the greatest urbanization rate at 82%, and 0.46 kg per person per day in Sub-Saharan Africa, which has a 38% urbanization rate. Waste production is likely to rise most in regions with a high number of developing low-income and lower-middle-income nations, as trash creation is typically expected to increase with economic development and population expansion [12]. Due to economic growth and urbanization, waste levels in Sub-Saharan Africa and South Asia are predicted to almost quadruple and double during the next three decades. Waste levels should grow more gradually in regions with higher-income countries, such as North America, Europe, and Central Asia [48].

3 Case Study Related to Uncollected Solid Waste

3.1 Accra, Ghana

Every year, Ghana undergoes serious seasonal flooding. Drains clogged with plastics and other debris are a major contributor in Accra, as they are in many other cities. Because of a lack of systematic collection, many people simply discard their solid waste, some of which is rinsed down the drain. Litter on the streets and gutters is also abundant, with much of it resulting from the indiscriminate disposal of commonly used plastic sachets containing drinking water. Floods in 2011 claimed lives, injured or destroyed livelihoods, and depleted economic value. Fourteen people were murdered, 43,000 were displaced, and 17,000 lost their homes, with roads, canals, and bridges all damaged. Furthermore, 100 cases of cholera were detected a week after the flooding occurred [30, 95].

3.2 Plague-Like Epidemic in Surat, India

Surat experienced a plague-like illness epidemic in 1994 as a result of heavy floods caused by uncollected garbage obstructing drains. In previous years, the city's population had grown faster than the infrastructure required to offer proper SWM services, which were the duty of the municipality [11]. At the time, municipal solid trash collection was limited to 40% of the city, leaving the remainder uncollected and frequently disposed of in sewers and aquatic bodies. The illness epidemic resulted in 693 cases and 56 fatalities. Furthermore, the government may have incurred up to USD 2 billion in economic losses, including roughly USD 420 million in missed export profits (for example, the UAE banned all cargo transshipments from India). The sickness epidemic came right before the Deepavali holiday, and over 45,000 tourists cancelled their vacations to India [95]. Over the next 18 months, Surat Municipal Corporation converted the city into one of the cleanest in the region. Slum infrastructure development, monitoring, private sector participation in waste collection and transportation, capacity building and coordination among municipal employees, public awareness raising, and the implementation of complaint handling systems were among the initiatives undertaken. They also established the Littering Detection Squad, an effort in which people labor to maintain the cleanliness of selected locations considered most vulnerable to litter and generate cash by penalizing citizens or shops who continue to litter despite earlier warnings [11, 95].

3.3 Love Canal, USA

Originally intended to be a communal settlement, Love Canal was to be powered by a canal that connected Niagara Falls and the Niagara River. By the time the project was abandoned, just one mile of the canal had been dug. After the land was purchased by Hooker Chemical Company, the canal was turned into a chemical and municipal waste site. Between the 1920s and the 1950s, some 20,000 tons of hazardous chemicals were disposed of there. When the canal could hold no more, it was filled up with dirt and sold in 1953 [62]. There, around the close of the 1950s, were houses and a school built. As the water table rose and tainted groundwater ascended to the surface, issues with odours and residues were documented in the 1960s and 1970s. Due to the decay of the waste drums, a record quantity of rainfall in 1978 caused polluted leachate carrying dangerous compounds to migrate into the basements and yards of the residences as well as the school, poisoning the air [5]. A significant public uproar resulted from the discovery of over 100 kid cases of diseases brought on by chemicals. Eleven of the 82 distinct chemicals discovered throughout the investigation are thought to be carcinogenic. A USD 10 million emergency financial relief package was agreed to relocate some of the people, and following the discovery of hazardous waste sites like Love Canal, the US implemented the Superfund law 14 in 1980. At these locations, cleanup efforts are still ongoing. The chemical business consented to pay the federal government back approximately USD 129 million in 1995 as part of a consent decree for prior expenditures related to response activities, claims for natural resource damages, and supervisory expenses [5, 95].

3.4 Naples, Italy

SWM has long been an issue in the Naples metropolitan region. The inability to develop a workable long-term solution for the treatment and disposal of garbage has resulted in recurring "crises" where waste could not be collected because there was nowhere for it to go. Waste that was allowed to accumulate in the streets served as a haven for vector-borne illnesses, endangering the public's health. There have occasionally been new, uncontrolled dumpsites used as emergency measures close to the metropolis, which has sparked more public outcry. These crises were widely publicized in 1994, 1999, 2003, 2008, 2010 and 2011 [51]. The fact that criminal organizations are actively involved in the transportation of rubbish has further complicated the problem and created significant tension among the many stakeholder groups. The trash management agency was headed by a prominent anti-waste activist, who was chosen by the new municipal government in 2011; waste was then shipped to other parts of Italy and the EU while a long-term local solution was being looked for [26, 40]. An incinerator with a 650,000 t/y processing capacity began operations at Accera, close to Naples, in 2013. Recycling of waste has also grown. The situation in the city has greatly improved as a result. Nevertheless, because the Naples region lacks composting facilities, trash is nevertheless carried to other areas, particularly organics, and is also shipped to the Netherlands for burning. What to deal with the 8 million tons of rubbish that accumulated at "temporary" sites in the vicinity of Naples remains an unresolved issue since it takes funding and political will, both of which are now in short supply [26, 95].

3.5 Barbados, Caribbean

One of the biggest sectors in the world, tourism contributes significantly to both national GDP and occupation in many countries. For instance, the Caribbean Island nation of Barbados welcomes over a million visitors annually, whose economic contribution accounts for about half of the country's GDP and jobs [84]. The spotless sea and coastal environment of this location draws a lot of people. Because of the negative ecological and aesthetic effects, the financial burden of governmental and private clean-up efforts associated with coastal and marine litter, and the connection between litter and a decline in traveller demand, marine and coastal litter is a problem that Barbados is particularly concerned about. According to a study conducted in Barbados with funding from The Caribbean Tourism Organization and The Ministry of Tourism of Barbados, beach litter may result in financial expenses due to its negative impacts on visitors' likelihood of returning [84, 95].

3.6 Payatas, Quezon City, Philippines

A slum neighborhood was devastated in July 2000 when a slope of the Payatas municipal dumpsite fell due to 10 days of intense rain. Rubbish pickers lived in the 18-ha complex, which took in 1500–1800 tons of rubbish a day on average. Their homes were built up to and onto the waste slope. Almost 300 individuals were killed in the avalanche, and hundreds more families were left homeless and without a means of support. Due to lack of soil cover, severe rain and high infiltration, the leachate level rose, causing the slope to break. Other contributing factors included landfill gas pressures, low waste density from minimal compaction, and unstable slopes at too sharp an angle. The garbage dumps had been shut down following the tragedy, but local officials were obliged to reopen it in November due to the impact on the city's cleanliness [95].

4 Good Practices for Solid Waste Management (SWM): Few Cases Studies

4.1 Waste to Energy Strategies in India

In India, the rate of material consumption has increased significantly over the years, with a six-fold rise from 1970 to 2015. Furthermore, projections indicate that this consumption is expected to double by 2030. The population of cities is projected to experience substantial growth. Due to the scale of waste generated, India will face significant operational challenges that demand substantial resources for effective management [28]. The Department of the Principal Scientific Advisor to the Government of India, established the Waste to Wealth Mission. It is a significant initiative aimed at utilizing global technological abilities to produce socio-economic advantages for the vast population of India. By tackling the challenges of waste disposal, declining air quality, and water pollution, this mission aims to benefit 1.3 billion Indians. The Principal Scientific Adviser (PSA), Government of India (GOI) leads the mission [39]. This mission is focused on leading scientific research and experimentation to develop and execute technologies that can effectively handle waste, recycle materials, generate energy, and extract valuable resources. Furthermore, it aims to create a thorough database of cutting-edge technologies from both national and international sources to assist Urban Local Bodies in tackling their waste management issues [39].

The primary projects involved in this initiative are the Agri-waste to Biocoal Pilot Plant in Mohali, Punjab, the Babarpur and Gokulpuri Drain Project in East Delhi, and the Waste Mining and Land Recovery of the Ghazipur Dump Site in East Delhi. In 2015, the Ministry of Environment, Forestry, and Climate Change created indicators for recycling and waste energy recovery under the heading 'Management and Trans-boundary Movement'. They introduced the subsequent guidelines: The introduction of environmentally conscious Standard Operating Procedures for Waste Management (SOP) is the first step towards (1) identifying locations for long-term storage and treatment; (2) disposing of hazardous wastes from various sources; (3) prioritizing waste co-processing for energy recovery over disposal; and (4) allowing hazardous waste imports from other nations for recycling, recovery, and reuse. These actions include rethinking the eco-labeling program to be more in line with circular goals, providing waste management training and capacity building for the unorganized sector, and allowing consumers to purchase products based on their environmental impact. Consequently, it is anticipated that this movement will have a favorable effect on the circularity flow in the waste management and energy sectors [73, 76].

4.2 Nairobi, Kenya—TakaTaka Solutions

TakaTaka Solutions is a vertically integrated waste management social enterprise based in Nairobi, Kenya. It works through a value-chain model in which it is actively involved in solid waste collection, sorting, recycling, composting, and disposal. It offers affordable SWM services to over 21,500 households and more than 40 corporates where it collects over 60 tonnes of waste daily. TakaTaka is creating jobs for many people who work in waste collecting, sorting, recycling, composting, and disposal activities. By 2023, the project had created up to 2400 jobs, up from just 200 in 2020 [86]. Waste is collected and transported using dump trucks and delivered to the entity's different plants. Organic waste (69%) is taken to the composting plant and composted into high-quality organic fertilizer which is sold at a profit to farmers. Plastic waste (26%) is recycled into new raw materials and sold to manufacturers. TakaTaka also partners with external recycling partners for other recyclables like paper, cardboard, glass, and many other materials. TakaTaka Solutions recycles 95% of all collected waste and disposes only 5% used in landfills [9, 34, 86].

4.3 Dhaka, Bangladesh—Waste Concern Waste Project

Waste concern started with a 'cash for trash' SWM model aimed at converting waste into resources in Dhaka. It aims to improve urban environmental sanitation through an SWM model that emphasizes waste reduction, reuse, recycling, and recovery. At the plants, organic is composted and converted into bio-fertilizer which is sold to farmers. The money earned is given back to help the communities. After a successful pilot project, the model attracted some funding partners resulting in the signing of public private-community partnership agreements between the communities, municipality, Bangladesh Ministry of Environment, and United Nations Development Programme (UNPD). Through funding from these partners, the model was replicated in five other communities across the city before attracting other municipalities [9, 44, 60].

4.4 Belo Horizonte, Brazil—Belo Horizonte Recycling Model

The city of Belo Horizonte has an integrated SWM system that promotes the separation of waste at the source to minimize the harmful waste-related environmental impacts and maximize socioeconomic benefits obtained from waste. The integrated SWM system is based on recycling, social inclusion (integrating informal workers into the formal SWM), job creation, and income generation. SWM in the city includes recycling construction waste and composting organics. All collected waste is taken to warehouses run by informal waste pickers' cooperatives where it is sorted, shredded, packaged, stored, and sold to industries. The money is shared among cooperative members. In managing construction waste, informal collectors collect debris from households for a fee (paid by households) using horse-drawn carts and deposit it at any of the 29 municipal receiving depots. The waste is used to make bricks for sale and some at municipal sanitary landfills. Organic waste including food, vegetables, and tree/plant material is sent to a composting facility and processed into compost manure which is sold to farmers [9, 17, 23].

4.5 Addis Ababa, Ethiopia—Reppie Waste-to-Energy Plant

Inaugurated in 2018, the Reppie plant is Africa's first waste-to-energy facility built after 116 waste pickers died in a landslide at the Koshe dumpsite. Addis Ababa City Administration undertook to rehabilitate the dumpsite and build an environmentally friendly plant that also generates electricity from waste. It received financial and technical assistance from the Ethiopian national government, UN-Habitat, the government of Japan, and private companies from Singapore, China, and Denmark. Koshe is an open dumpsite where residents can dump waste for free. Once delivered, waste is sorted and separated according to type. The plant is divided into a composting plant (for organic waste) and an incinerator (for burnable waste). Organic waste is decomposed using the Japanese 'Fukuoka method' which speeds up decomposition and reduces the generation of greenhouse gases like methane which contributes to global warming [2]. It is converted into high-quality bio-fertilisers and sold to farmers. Burnable inorganic waste is burnt in the plant's combustion chamber to boil water to produce steam which is used to power a turbine generator to produce electricity. The plant covers over 350,000 tonnes of waste into 50 MW/year and supplies 30% of Addis Ababa's household electricity. Informal waste pickers also earn income from picking waste for reuse, recycling, and selling it to industries. SWM activities at the Reppie plant reduced odour, pollution, erosion, flooding, and disease outbreaks in nearby neighbourhoods. It created jobs for plant workers and improved livelihoods for informal waste pickers. The model received attention from Kenya and the Nairobi County is set to construct a similar plant with funding from the African Development Bank. However, the Reppie plant faces challenges related to inadequate financing, low quality waste, and lack of skilled personnel, which reduce its capacity to fully function [9, 80, 98].

4.6 Nairobi, Kenya—Mukuru Integrated Waste Recycling Centre

Located near Nairobi's Dandora dumpsite, Mukuru Recycling Center is a community-driven project that uses garbage to generate livelihoods. The Kenya Association of Manufacturers, Nairobi City Council, UN-Habitat, and St. John's Catholic

Church in Korogocho collaborated to give both financial and technical support once it was established. The project's main goal was to reduce the exploitation of garbage pickers by local waste dealers, rehabilitate those who stayed at the dumpsite, and turn rubbish picking into a sustainable source of income [9, 41]. There are four main types or categories into which SWM activities are divided. Group 1 removes recyclable rubbish from the landfill, classifies and cleans it, and then sells it straight to businesses, cutting out middlemen. Group 2 is paid to collect wastepaper from commercial establishments in the CBD and resells it to recycling enterprises directly for a profit. Group 3 is a company that provides waste collecting services. Group 4 works on the organic garbage that is brought to the landfill. In addition to using, it for personal domestic use at home, they use it to manufacture stock feed for pigs and bio-fertilizer that they sell to farmers. The team also handles plastic garbage, which they gather and transport to a recycling facility where it is sorted by colour, crushed into pellets, and sold to businesses. Group 4 uses sawdust, coffee husks, and/or low-quality waste papers to manufacture briquettes that are sold. The Mukuru initiative has not only helped homeless garbage pickers reintegrate into society and provide employment and livelihoods, but it has also decreased crime, disease outbreaks, and flash floods in the neighbourhood [9, 41].

5 Indian Scenario of Current Waste Generation Scenario

India is presently one of the top 10 nations in the world in terms of municipal solid waste generation. According to the most recent solid waste data from the Central Pollution Control Board (CPCB), the country generates 160.038 TPD (tonnes per day). The overall collection efficiency is 95.4%, with 152.7495 kTPD of trash collected. 79.9563 kTPD is processed, while 29.4272 kTPD is disposed of in landfills [66]. Nonetheless, there is a significant difference of 50.6554 kilotons per day (kTPD), or 31.7% of total trash created, which remains unaccounted for. Delhi has the most solid waste generation per capita in India, followed by Lakshadweep and Mizoram. Furthermore, the percentage of processed solid waste has been increasing, with the amount expected to climb from 19% in 2015–16 to 49.96% by 2020–21 [16]. According to the most recent data from the Central Board of Pollution Control, Maharashtra, Uttar Pradesh, and Tamil Nadu generate the most solid rubbish in India. Chhattisgarh is the only state in India where the quantity of rubbish produced equals the amount treated. In terms of total waste treatment, Maharashtra leads, followed by Tamil Nadu and Gujarat. Over the past six years, the amount of rubbish shipped to landfills has steadily decreased. From 54% in 2015–16 to 18% in 2020–21, landfill disposal of solid waste has substantially reduced. The majority of India's garbage disposal sites (221) are located in Maharashtra (382), Madhya Pradesh (341), and Karnataka (275). Furthermore, Maharashtra and Karnataka have the most active landfills in India, with 137 and 52 sites, respectively [66].

6 Government Initiative for Management of Solid Waste in India

India is a country that is comprised of a vast variety of geographical regions, which results in a varied range of rubbish creation patterns among its diversified population. Additionally, there is still a dearth of real programs that are intended to investigate the patterns of waste generation in certain locations and areas. Over the course of the past ten years, there has been a substantial expand in the pace of urbanization over the total population of the nation. The huge increase in population has led to an increase in the amount of municipal solid waste that is being produced [64]. As per the SWM Rules 2016, the Ministry of Housing and Urban Development has issued a revised edition of the Municipal SWM Guidelines 2016 to aid states and cities in effectively managing solid waste. In addition to providing suggestions for urban local authorities, it helps states and cities make decisions that are beneficial for the sanitary and waste management system, all in an effort to achieve the Swachh Bharat Mission's objectives. In India's history of sanitation, the Swachh Bharat Mission-Grameen (SBM-G) symbolizes a revolutionary period. It is driven by a past that is enriched with ancient innovations and is supported by contemporary efforts made by the government. Formed in 2014 by the Honourable Prime Minister, the mission's primary objective was to eliminate the practice of open defecation (ODF) throughout India.

By effectively rallying the entire country to participate, this massive undertaking has solidified its position as the most impactful global movement for behavioural change. Fitting with Sustainable Development Goal Target 6.2, it was a commemoration of Mahatma Gandhi's 150th birthday. In 2019, the program saw the completion of over 100 million individual household toilets, and over 6 lakh villages were designated to be part of the open development framework [43]. Within the context of the Swachh Bharat program, SWM is an essential tool. For a period of five years, the central government gives financial assistance to urban regions for the purpose of SWM. A greater number of expectations are placed on the state governments and the urban local bodies (ULBs) to concentrate on establishing a priority, effectively carrying out the plan, and providing the conclusion that will assist in defining the scope of the mission and ensuring compliance with it. In India, environmental policies and activities are developed, coordinated, and monitored by the Ministry of Environment, Forest, and Climate Change, an important government institution. An important objective of the Ministry of Environment, Forests, and Climate Change (MoEFCC) is to control and prevent pollution. Minister of State for the Ministry of Environment, Forests, and Climate Change (MoEFCC) reports that just six states have managed to produce inventories of electronic waste generation; Jammu & Kashmir is one of these states. In the Indian Parliament's Lok Sabha (House of the People) session, this data was presented in answer to question 4553. In addition to the states of Goa, Jammu & Kashmir, Punjab, Madhya Pradesh, Chhattisgarh, and Himachal Pradesh [10].

In agreement with the E-waste (management) laws of 2016, the state government has given permission to the industry department or any other government body to allot space inside the industrial park for the purpose of dismantling and recycling electronic trash. This license was provided by the state government. The following points will be executed and ensured by the state labour department, which will be accountable for their implementation [78]. It is mandatory to register workers who are involved in dismantling and recycling operations. Assist in the organization of labour organizations to set up dismantling facilities. It is critical to participate in skill development initiatives to guarantee the well-being of workers performing dismantling and recycling tasks. In order to prioritize the safety and well-being of workers, it is imperative that they undergo routine health monitoring. An annual report must be submitted by the state government to the Civil Commission and the Ministry of Environment and Forestry outlining the strategy to implement all of these regulations [54].

The Government of India has released a notification relating to the bio-medical waste (BMW) rule, establishing a comprehensive framework for managing the waste produced within the country. 2016 BMW management guidelines include a wide variety of essential areas, including the following: Preparing the garbage from the laboratory, the waste from the microbiological samples, the blood bags, and the blood samples. It is important to provide all healthcare personnel with complete training on the legislation and procedures that pertain to the treatment of biological waste. Vaccinations against hepatitis-B and tetanus should also be obtained by everyone entering the healthcare field. The goal is to put barcoding and GPS technology on waste carrying vehicles manufactured by BMW within a timescale of one year. For a period of five years, it is imperative that records of incineration, hydroclaving, and autoclaving procedures be maintained. BMW can be broken down into four distinct groups, each of which is structured according to a different treatment strategy. In the event that untreated human anatomical waste, animal anatomical waste, dirty garbage, and biotechnological waste are stored for a period of time exceeding forty-eight hours, authorization is not something that is required [19].

7 Evaluating Major Reason Behind Solid Waste Generation

By the year 2050, it is projected that there will be a staggering 3.40 billion metric tons of garbage generated annually. This signifies a 70% rise in waste production compared to the previous year. A notable factor contributing to this increase is the simultaneous rise in urbanization rates and global population. According to scientific projections, the quantity of waste generated per individual is projected to rise by 19% in high-income countries by 2050, while low and moderate-income countries are anticipated to experience a minimum increase of 40% [66].

A lot of factors make municipal solid waste (MSW) management a difficult and complicated task. Management, structure, and complexity may be unnoticed due to inadequate resource allocation. Municipal solid waste (MSW) can differ greatly from

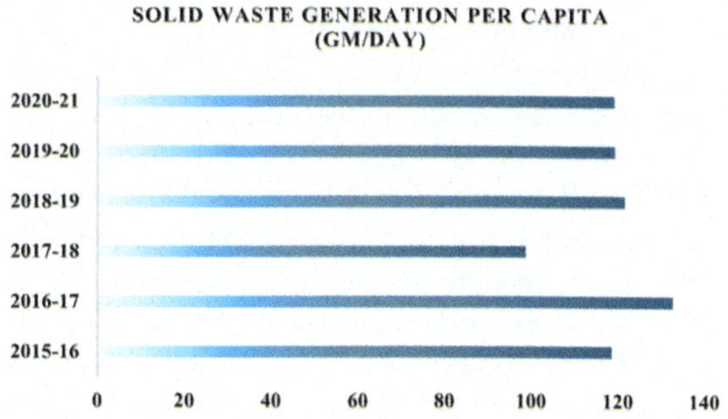

Fig. 3 Waste generation per capita

one nation or city to another for many reasons. Some such reasons includes economic climate, population density, industrial makeup and waste-management regulations. In order to evaluate the likelihood of using MSW as a power source, engineers and scientists need a thorough knowledge of the waste's energy content and composition. Gasification and incineration are two important energy conversion processes used to handle municipal solid waste, and this is helpful for forecasting the makeup of gas emissions from these processes. These numbers will be essential for the enhancement of systems that can effectively produce electricity from municipal waste. Effective waste management significantly aids in keeping the informal economy afloat [24].

Around 70% of waste can be recycled according to the reports of scientific studies. Only 30% includes the non-recyclable portion such as soil and leftover food. A full understanding of these factors is necessary for advancing the making of successful waste management strategies and the introduction of ecologically conscious activities [24]. Figure 3 displays the waste production per capita throughout the years. Those with less than one lakh inhabitants produce solid waste at a rate of 0.36 kg per capita per day, whereas those with in excess of two crores of inhabitants produce solid waste at a rate of 0.43 kg per capita per day. The population size is one of the most significant things to look at if you want to know the amount of solid waste a country produces [83].

Historical studies have shown that garbage collection, sorting, transportation, and treatment present substantial difficulties in India's urban areas. In addition, many developing nations struggle with the problems that result from SWM. Problems with waste treatment technology, limited financial resources, insufficient knowledge and effort, inappropriate waste-related actions, and similar challenges are among them. There is a robust relationship between community social activities and solid waste amounts. The composition and amount of solid waste generated by homes are affected by several other factors as well. Modifications to commodity consumption and source-sorting behaviour are two of these factors. The available resources present the greatest worldwide challenge to municipal SWM, alongside economic, sociological, legal, political, and environmental considerations. This means that the community's social, cultural, and economic health should be considered alongside any potential new technology for MSW management [77, 83].

8 Impacts of Solid Waste on Humans and the Environment

Solid waste management (SWM) in developing countries is plagued by a number of constraints, including those of and technological, regulatory, economic, institutional, and operational nature [38]. Garbage collection and disposal in landfills or dumpsites is standard practice in most developing countries. Most of these facilities are expected to achieve capacity over the next decade. Tossing trash into bodies of water or dumping or burning it in open areas, which were often in low-income districts on the city's outskirts, were both seen as acceptable methods of waste disposal. It is not sustainable to dispose of rubbish in this way. The use of outdated or badly maintained facilities, unregulated informal dumping, or open-air trash burning is also practised by several localities. People from previously marginalized communities are often the ones most affected by these activities in the areas surrounding landfills [4]. The major impacts caused by solid waste and its improper handling is explained in Fig. 4.

Further, the commencement of the environmental movement in the 1960s marks the beginning of broad realization of the hazards that are presented to public health and the environment by SWM practices that are not sustainable. As a result of the perception that SWM was a technological issue that could be solved through technological means, the equipment used for garbage collection gained the greatest attention and investment beginning in the 1970s and continuing to this day. Even though technology had made significant progress in lowering emissions of hazardous substances, this perspective reversed in the 1990s when cities realized they required assistance from citizens and companies to properly remove and dispose of waste. This was the case despite the fact that technology had developed [89]. The problems mentioned above lead to the accumulation of many sorts of waste, such as plastic bags, plastic bottles, paper scraps, paper-plastic containers, and decomposing trash in streets and

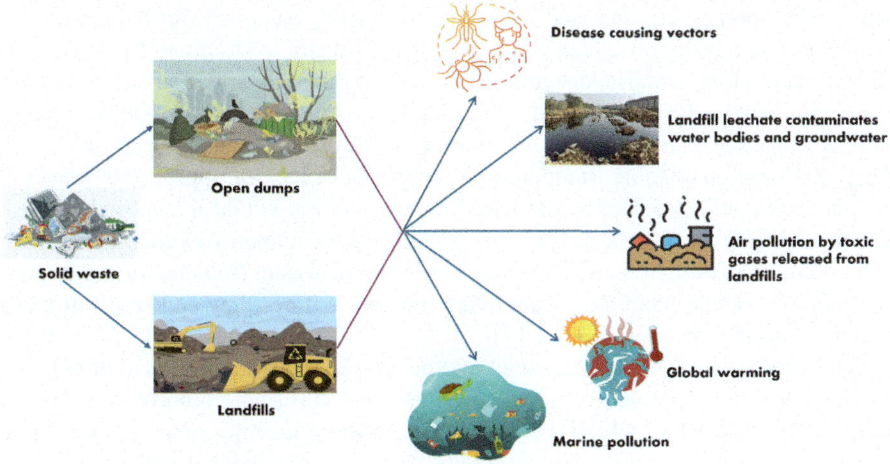

Fig. 4 Solid waste major impacts

drains. The lack of essential services, such as waste collection, sanitation facilities, and sanitary water delivery, is making the problem worse [66].

Additionally, the World Health Organization (WHO) has highlighted the negative consequences of inappropriate solid waste disposal, incorporating pollution of land, water, and air, and the health effects on nearby populations. The potential negative effects of solid waste on human health is influenced by many factors. These include waste management systems, the nature and habits of the exposed population, the duration of exposure, and measures to reduce or eliminate the issue. The landfills which are not lined causes very serious threat to the environment in developing nations due to leakage of heavy metals and various hazardous compounds. Waste management practices that dispose solid waste into or near water sources degrade water quality to very high extent [29].

The negative influences on the society and environment that are caused by waste that is not collected and is handled improperly are felt far beyond the boundaries of any one city. The procedure has an array of negative impacts on the sustainability of the environment, including the emission of methane (CH_4), the production of a disagreeable stench, the pollution of the air, the contamination of the water and the land, and the expansion of insects that transmit diseases, flies, and rodents. The breakdown of biodegradable waste in anaerobic environments is responsible for approximately 18% of the globe's methane emissions and 2.9% of its greenhouse gas emissions with an influence on global warming that is approximately 25 times greater than the emissions of carbon dioxide (CO_2) [3, 8]. Inconsistent garbage collection is another factor that contributes to contamination in maritime environments. As of the year 2010, 192 coastal nations were responsible for the generation of 275 million metric tons of waste plastic, of which up to 12.7 million metric tons (4.4%) was introduced into ocean ecosystems [42].

The present environmental and climatic changes, which are happening on a global scale and directly affect health and wellbeing, are now generally believed to be caused by human activities and their outcomes [94]. Also, several human actions lead to garbage being produced at the local level. Infectious diseases like respiratory disorders, cholera, malaria, dysentery, and injuries are only a few of the many environmental and health hazards that these pollutants exacerbate. More people living in cities means more solid waste, which is bad for the environment and people's health [1, 75]. Numerous factors, including the existence or lack of appropriate preventative and mitigating measures, the length of time an individual is exposed to trash, the type of rubbish, and the community, are among the numerous variables that have the potential to influence the health effects of solid waste. Sometimes the impacts on people's mental health are very minor, but other times they can result in major sickness, disability, or even death [7].

The formation of landfill leachate another of the most deteriorating effects that landfills, and improper waste management can have. It is a dark colored liquid which emits unpleasant odour which is collected below waste trenches or a leachate collection pond [67]. It is a liquid that percolates from the waste because of precipitation. This wastewater is extremely hazardous since it contains a variety of hazardous components that have an impact not only on the health of humans but also on the environment. The amount of waste generated by a city and the number of people living there may be proportional to the amount of pollution caused by landfill leachate in that city. As a result of the lack of a liner system in the leachate pond, incorrect treatment, and surface runoff caused by monsoon floods, it has been discovered that the leachate extracted from these landfills has contaminated both the surface water and the ground water [65].

An open dumpsite in Iraq was the subject of recent research that investigated how seasonal shifts in leachate from the dumpsite affected the quality of groundwater. The key characteristics of the leachates were their toxicity, which was caused by the existence of heavy metal concentrations, in addition the excessive levels of organic and inorganic components that they included together. Nearly every physicochemical parameter is found in significantly higher concentrations in the leachate emanating from the dump site [74]. One of the most significant causes of water contamination is the disposal of solid waste into water systems. The presence or absence of landfills determines the water quality. The riverbank residents are another group that feels the effects. Scientific management and the identification and removal of pollution sources can stop water contamination in its tracks [82].

Research has shown that heavy metals present in the hazardous leachate can cause cancer. It was determined that cadmium, with its high chromium values across all landfills, is the most potent carcinogen, followed by lead, chromium, and nickel, according to the risk assessment [14]. Physiochemical parameters, anions, and cations were found to be more than the permissible limit in the leachate sample that was analyzed in a different investigation. Most of the samples, which were found in close proximity to the landfill sites in Mohali and Chandigarh, contained exceptionally high concentrations of ammoniacal nitrogen and chloride. The fact that this is the case indicates that there is some pollution of the groundwater by leachate from the landfill [61].

In open dumps that contain organic garbage, pests that carry disease, such as flies, mosquitoes, and rats, thrive due to the presence of organic waste. Dengue fever, Zika virus, and malaria fever are all diseases that are transmitted by vectors and are frequently connected with one another [20, 32, 35, 57]. There is also the chance of contracting water-borne infections such as leptospirosis, hepatitis A, intestinal worms, diarrhoea, and other similar conditions [63, 68].

9 Remediation Techniques and Disposal Strategies for Waste

In India, where over a thousand dumpsites have been documented, around 79% of the collected waste ends up in these landfills. Some of the problems linked to the ancient dumpsites include the release of harmful gases and the migration of leachate into groundwater [55]. Concerns like these have the potential to harm both the local ecology and the health of the people who reside there. Also, landfill fires and slope failures are on the rise as issues with disposal sites [81]. Different nations, both developed and developing, have long held different opinions regarding the most effective way to deal with solid waste. It is important to keep in mind that the population and gross domestic product of a country have a considerable impact on the efficiency with which its municipal solid waste is managed, whether through recycling, diversion, or conventional disposal. As a result of the development of waste-to-energy technology, the process of transforming municipal solid waste into fuels, heat, and power has become a well-established practice in manufacturing nations [59].

It is becoming increasingly difficult to find enclosed sanitary dumping facilities, which is one reason why dump sites in major cities are rapidly getting overcrowded. In addition to this, it is essential to incorporate the appropriate disposal of hazardous waste, which includes trash from electronic devices, batteries, and sanitary products, into distinct disposal processes. A level of SWM dumping in India has reached an unbearable level. Facilities that are able to manage and dispose of increasing quantities of municipal solid waste are therefore required [66].

The process of recycling, anaerobic digestion, and composting only account for a quarter of the world's collected municipal solid waste. Since different countries have different approaches to recycling, it is critical to identify the best way to dispose of or handle municipal solid waste on a worldwide scale [58]. Burying non-recyclable garbage is a typical practice all throughout the world. But instead of burying garbage with dirt, some underdeveloped countries just pile it high or hurl it into pits. One common way that industrialized nations dispose of their municipal solid waste is by using landfills. However, in poorer nations, where high population density makes limited open space a concern, it is not as widely practiced. In developing nations, some landfills double as temporary trash storage locations and have facilities for sorting, recycling, and treatment in addition to waste consolidation, transfer, and processing. It is possible to monitor the processing of garbage and separate recyclables before tipping them in a landfill, making it an ideal location for municipal solid waste disposal [58].

Reclamation and landfill mining are also a unique method to improve material recycling and energy recovery from closed landfills. It examines the advantages, disadvantages, and uncertainties associated with this strategy. In the year 1953, landfill mining began as a way to retrieve decomposed waste from the Hiriya Landfill located near Tel Aviv, Israel. The main objective was to reclaim these byproducts and utilize them as fertilizers in agricultural environments, specifically in orchards [92]. In recent years, the interest of landfill owners in landfill mining has been fueled by the increase in prices of commodities, as well as the possibility of generating revenue through the recycling of dig out rubbish and the development of energy. Valuable recyclable items, compost and combustible waste gases are recovered as a process of landfill mining which further results in reduction in footprint of the landfill [46].

One noteworthy characteristic of landfill mining is decrease in overall landfill mass contained within the closed dump. A crucial step in making space for new waste for disposal is digging out organic material, whether partially or totally decomposed as well as items which are recyclable. Two main types of landfill mining processes are in-situ ad ex-situ mining [46]. In situ landfill mining involves utilizing resource

54. MoEFCC (2016) E-waste (management) rules. Government of India, Ministry of Environment, Forest and Climate Change, India
55. Mor S, Ravindra K (2023) Municipal solid waste landfills in lower- and middle-income countries: environmental impacts, challenges and sustainable management practices. Process Saf Environ Prot 174:510–530
56. Moya D, Aldás C, López G, Kaparaju P (2017) Municipal solid waste as a valuable renewable energy resource: a worldwide opportunity of energy recovery by using waste-to-energy technologies. Energy Procedia 134:286–295
57. Muiruri J, Wahome R, Karatu K (2020) Assessment of methods practiced in the disposal of solid waste in Eastleigh Nairobi County, Kenya. AIMS Environ Sci 7:434–448
58. Nanda S, Berruti F (2021) Municipal solid waste management and landfilling technologies: a review. Environ Chem Lett 19:1433–1456
59. Nanda S, Berruti F (2021) Thermochemical conversion of plastic waste to fuels: a review. Environ Chem Lett 19:123–148
60. Nasrin ST (2014) Urban development on municipal solid waste management in Dhaka, Bangladesh. Degree of master—environmental engineering and sustainable infrastructure, KTH Royal Institute of Technology, University in Stockholm, Sweden
61. Negi P, Mor S, Ravindra K (2020) Impact of landfill leachate on the groundwater quality in three cities of North India and health risk assessment. Environ Dev Sustain 22:1455–1474
62. Newman RS (2016) Love canal: a toxic history from colonial times to the present. Oxford University Press
63. Olay-Romero E, Turcott-Cervantes DE, Hernández-Berriel MDC, Lobo-García De Cortázar A, Cuartas-Hernández M, De La Rosa-Gómez I (2020) Technical indicators to improve municipal solid waste management in developing countries: a case in Mexico. Waste Manage 107:201–210
64. Pal MS, Bhatia M (2022) Current status, topographical constraints, and implementation strategy of municipal solid waste in India: a review. Arab J Geosci 15:1176
65. Parvin F, Tareq SM (2021) Impact of landfill leachate contamination on surface and groundwater of Bangladesh: a systematic review and possible public health risks assessment. Appl Water Sci 11:100
66. Patel J, Mujumdar S, Srivastava VK (2023) Municipal solid waste management in India—current status, management practices, models, impacts, limitations, and challenges in future. Adv Environ Res 12:95–111
67. Patel J, Mujumdar S, Srivastava VK (2024) Status of landfilling in developing and developed nations: a literature review. In: Solid waste management for resource-efficient systems, pp 273–299
68. Penteado CSG, Castro MASD (2021) COVID-19 effects on municipal solid waste management: what can effectively be done in the Brazilian scenario? Resour Conserv Recycl 164:105152
69. Permana AS, Towolioe S, Aziz NA, Ho CS (2015) Sustainable solid waste management practices and perceived cleanliness in a low income city. Habitat Int 49:197–205
70. Phan LT, Nguyen GT, Nguyen QAD, Nguyen HS, Nguyen TT, Watanabe T (2021) Quality of life and factors affecting it: a study among people living near a solid waste management facility. Front Public Health 9
71. Pomoni DI, Koukou MK, Vrachopoulos MG, Vasiliadis L (2024) Circular economy: a multilevel approach for natural resources and wastes under an agri-food perspective. Water-Energy Nexus 7:103–123
72. Pongrácz E (2002) Re-defining the concepts of waste and waste management: evolving the theory of waste management
73. Priyadarshini P, Abhilash PC (2020) Circular economy practices within energy and waste management sectors of India: a meta-analysis. Bioresour Technol 304:123018
74. Rashid SW, Shwan DMS, Rashid KA (2022) Physicochemical characterization and evaluation of seasonal variations of landfill leachate and groundwater quality around Tanjaro open dump area of Sulaymaniyah City, Kurdistan, Iraq. J Chem 2022:8574935
75. Rauf M, Saleem MD, Anwer MO, Ahmed G, Aziz S, Memon MA (2013) HIV, hepatitis B and hepatitis C in garbage scavengers of Karachi. J Pak Med Assoc (JPMA) 63:798–802

76. Rezania S, Oryani B, Nasrollahi VR, Darajeh N, Lotfi Ghahroud M, Mehranzamir K (2023) Review on waste-to-energy approaches toward a circular economy in developed and developing countries. Processes 11
77. Sahoo A, Dwivedi A, Madheshiya P, Kumar U, Sharma RK, Tiwari S (2024) Insights into the management of food waste in developing countries: with special reference to India. Environ Sci Pollut Res 31:17887–17913
78. Sengupta D, Ilankoon IMSK, Dean Kang K, Nan Chong M (2022) Circular economy and household e-waste management in India: integration of formal and informal sectors. Miner Eng 184:107661
79. Sheavly SB, Register KM (2007) Marine debris & plastics: environmental concerns, sources, impacts and solutions. J Polym Environ 15:301–305
80. Simegn A, Abebe S, Worku A (2021) Characterization and optimization of incinerated municipal solid waste fly ash as a cement substitute material in concrete at Reppie waste to energy plant in Ethiopia, East Africa. Adv Environ Stud 5:382–393
81. Singh A, Chandel MK (2020) Effect of ageing on waste characteristics excavated from an Indian dumpsite and its potential valorisation. Process Saf Environ Prot 134:24–35
82. Singh CR, Dey M (2014) Surface water quality with respect to municipal solid waste disposal within the Imphal municipality area, Manipur. Int J Sci Res Publ 4:1–4
83. Soni A, Das PK, Kumar P (2023) A review on the municipal solid waste management status, challenges and potential for the future Indian cities. Environ Dev Sustain 25:13755–13803
84. Spencer A (2018) Travel and tourism in the Caribbean: challenges and opportunities for small island developing states. Palgrave Macmillan, Cham
85. Sun L, Fujii M, Tasaki T, Dong H, Ohnishi S (2018) Improving waste to energy rate by promoting an integrated municipal solid-waste management system. Resour Conserv Recycl 136:289–296
86. Takataka-Solution. How we sort waste [Online]. Available: https://takatakasolutions.com/waste-recovery/. Accessed 12 June 2024
87. Tammemagi HY (1999) The waste crisis: landfills, incinerators, and the search for a sustainable future. Oxford University Press
88. Temmerman L, Veeckman C, Tsoumani O (2019) D4.1: setting the scope of the behavioural change campaign: a behavioural mapping exercise
89. Un-Habitat (2010) Solid waste management in the world's cities: water and sanitation in the world's cities 2010. Routledge
90. Urme SA, Radia MA, Alam R, Chowdhury MU, Hasan S, Ahmed S, Sara HH, Islam MS, Jerin DT, Hema PS, Rahman M, Islam AKMM, Hasan MT, Quayyum Z (2021) Dhaka landfill waste practices: addressing urban pollution and health hazards. Build Cities 700–716
91. Vähäsöyrinki HJ (2021) Circular economy and industrial packaging material waste in EU: case study
92. Van Der Zee DJ, Achterkamp MC, De Visser BJ (2004) Assessing the market opportunities of landfill mining. Waste Manage 24:795–804
93. Voukkali I, Papamichael I, Loizia P, Zorpas AA (2024) Urbanization and solid waste production: prospects and challenges. Environ Sci Pollut Res 31:17678–17689
94. Whitmee S, Haines A, Beyrer C, Boltz F, Capon AG, De Souza Dias BF, Ezeh A, Frumkin H, Gong P, Head P (2015) Safeguarding human health in the Anthropocene epoch: report of The Rockefeller Foundation–Lancet Commission on planetary health. The Lancet 386:1973–2028
95. Wilson DC, Rodic L, Modak P, Soos R, Carpintero A, Velis K, Iyer M, Simonett O (2015) Global waste management outlook. United Nations Environment Programme, International Solid Waste Association
96. Wynne B (1987) Risk management and hazardous waste: implementation and the dialectics of credibility. Springer-Verlag

97. Yaashikaa PR, Senthil Kumar P, Varjani SJ, Saravanan A (2019) Advances in production and application of biochar from lignocellulosic feedstocks for remediation of environmental pollutants. Bioresour Technol 292:122030
98. Yimam YM, Tesfaye N (2020) Environmental and health impacts of and challenges on Rappie Koshe waste-to-energy facility, Addis Ababa. Int Res J Multidiscip Scope 1:9–17

Sustainability in Focus: Exploring the Link Between Financial and Environmental Education Among Management Students

Suellen Caroline Bertulino Cavalcanti, Kliver Lamarthine Alves Confessor, Kettrin Farias Bem Maracajá, and André Cavalcante da Silva Batalhão

Abstract This study aims to investigate the relationship between individuals' perceptions of Financial Education and Environmental Education. It is a quantitative, descriptive research, with data collection carried out through the instrument developed by Couto et al. (Adm Ensino Pesquisa 23(3):510–534, 2022 [12]). The sample will consist of undergraduate students from the Federal University of Campina Grande (UFCG), Brazil, with the aim of analysing the correlation and behaviour of the variables within the sample. Considering that the topic of sustainability has been garnering the interest of numerous researchers and professionals from private and governmental organisations, and the scarcity of studies exploring the relationship between sustainable development and financial education, especially in its environmental dimension, the results of this research contribute to encouraging educational policies that promote the integration of financial and environmental dimensions. This aims to prepare future leaders to face the challenges of economic and environmental sustainability. Such integration will benefit not only the students but also have a positive impact on organisations and society, promoting a more balanced and sustainable future.

Keywords Financial education · Sustainability · Environmental dimension

---

S. C. B. Cavalcanti · K. L. A. Confessor · K. F. B. Maracajá
Department of Management, Federal University of Campina Grande, Campina Grande, Paraíba, Brazil

A. C. da Silva Batalhão
Department of Applied Management Sciences, Minas Gerais State University, Passos, Minas Gerais, Brazil

A. C. da Silva Batalhão (✉)
Graduate Programme in Environmental Sciences, Federal University of Goiás, Goiânia, Goiás, Brazil
e-mail: andrebatalhao@gmail.com

A. Cavalcante da Silva Batalhão and P. Singh (eds.), *Blending Approaches Towards Sustainability Focus*, Earth and Environmental Sciences Library,
https://doi.org/10.1007/978-3-031-93518-3_8

1 Introduction

With the drafting of the Brundtland Report during the World Commission on Environment and Development, the concept of sustainable development emerged, defined as the ability to meet the needs of the present generation without compromising the ability of future generations to meet their own needs Brundtland [6]. Later, with the publication of the paper "Towards the sustainable corporation: Win-win-win business strategies for sustainable development" by Elkington [15], sustainable development began to be evaluated according to three dimensions: environmental, social, and economic. These dimensions must be balanced to avoid social problems, economic dysfunctions, and environmental catastrophes [23].

One of humanity's greatest challenges is reconciling economic growth and environmental preservation [19]. Sustainable development can be characterised as a set of policies that ensure national income and access to basic rights such as economic security, health, and education, while minimising the impact of production and consumption on the environment [35].

For society to develop sustainably, it is essential that individuals perceive themselves as citizens and agents of transformation. Education for sustainability fosters the formation of well-informed citizens, with new values, skills, attitudes, and behaviours that promote harmony between social and ecological processes [42]. Learning becomes a transformative tool that generates critical thinking and collective actions that challenge normalised aspects of society, essential for the emergence of sustainability-related transformations [25].

Environmental education can be emancipatory by consolidating educational objectives linked to knowledge and social participation [5]. Through environmental ethics, environmental education can politicise individuals, transforming them into consumers who support sustainable practices [37]. Awareness of sustainable economic practices is essential for environmental preservation, and a healthy environment is essential for the long-term economy.

The critical aspect of financial education seeks a humanistic approach, strengthening individual capacities and citizen empowerment [8]. Financial education helps individuals in resource management, decision-making, and conscious consumption. Lack of financial knowledge can exacerbate problems such as poverty, environmental degradation, and climate change [38]. Discussing financial education from the perspective of sustainability and understanding how these two areas can complement each other contributes to the continuous improvement of society and to economic and environmental balance [12].

The relationship between financial education and environmental education is multifaceted and interconnected by principles and practices that promote individual and collective well-being, as well as economic and environmental sustainability. Therefore, this study seeks to answer the following question: What is the relationship between individuals' perceptions of financial education and environmental education?

This study aims to investigate the relationship between individuals' perceptions of financial education and environmental education. Specifically, it applies the instrument developed by Couto et al. [12] to a sample of undergraduate students in Business Administration from the Federal University of Campina Grande (UFCG), Brazil. This instrument captures the respondents' perceptions of the dimensions of financial education and environmental education. Subsequently, the correlation and behaviour of these variables are analysed within the sample context.

Considering the growing interest of researchers and professionals in sustainability and the scarcity of studies exploring the relationship between sustainability and financial education, especially in its environmental dimension, this study aims to fill this gap [12]. The results obtained can assist educational institutions and governments in promoting integrated environmental and financial education training and activities, forming more conscious individuals prepared for the development of a sustainable society.

2 Theoretical Backgrounds

2.1 Environmental Dimension of Sustainability

Since the dawn of humanity, the pursuit of development in various aspects has led to the exploitation of the environment, resulting in significant impacts. In agriculture, for example, the increasing use of pesticides has contributed to human intoxication, decreased soil fertility, and environmental contamination, leading to a process of socio-environmental illness [29]. With the Industrial Revolution in the eighteenth century, the adoption of a linear production scheme based on the extraction, production, commercialization, use, and disposal of products increased the exploitation of natural resources. This process, coupled with consumerism and waste disposal, caused resource scarcity, pollution, and environmental degradation [33].

Discussions on socio-environmental responsibility began only in the 1960s in the United States, in a context of poverty, unemployment, and increasing pollution. Socio-environmental responsibility is related to business decisions based on ethical values, incorporating legal dimensions, and respecting people, communities, and the environment [18].

Sustainability is not limited to the environmental context and preservation. Sustainable consumption, linked to the practice of conscious consumption, integrates economic, political, and social actions [2]. Cabral [7] reveals that unplanned consumption can lead to problems such as financial difficulties, increased inflation, and indebtedness, causing financial imbalance.

In 1972, at the United Nations World Conference on the Human Environment in Stockholm, the first significant action for awareness and the development of measures in this context took place. In 1987, with the report "Our Common Future," the concept of sustainable development was established as one that meets the needs of the present without compromising future generations [6]. In 1999, the publication of the book "Cannibals with Forks: The Triple Bottom Line" by John Elkington introduced the concept of sustainable development as economic prosperity, environmental quality, and social justice [16], presenting the three dimensions of sustainability: economic (profit), environmental (planet), and social (people).

Agenda 21, established in 1999 in Rio de Janeiro, reinforced the perception that environmental sustainability is linked to sustainable consumption and production models. To achieve sustainable development, the behaviour of those who supply and demand plays a decisive role, requiring conscious decisions. Education, especially Financial Education, plays a crucial role in promoting new consumption behaviours [10] and supply.

Understanding the environmental dimension of sustainability reveals the complexity of interactions between humans and the environment throughout history [13]. From the early days to the present, consumption and production practices have significantly shaped environmental balance [1]. Awareness of environmental preservation and the adoption of sustainable consumption and production models are imperative to achieving balanced development [3]. In this context, Financial Education emerges as a powerful tool to promote not only economic stability but also essential behavioural changes for a more conscious and sustainable society [11]. Recognising the interdependence between environmental sustainability and financial education allows a holistic approach to addressing contemporary socio-environmental challenges [12].

In higher education institutions, the ideology of efficiency has been an obstacle to critical education, reflected in the fragmented organisation of disciplinary knowledge [41]. In this scenario, Environmental Education needs an epistemological link that contributes to its inclusion [36]. Additionally, there is a shortage of professionals capable of developing and implementing effective environmental policies, as well as a lack of interdisciplinarity and environmentalisation in university curricula and research [39].

2.2 Financial Education and Its Economic-Environmental Direction

According to the Organisation for Economic Co-operation and Development [28], financial education is the process by which consumers and investors improve their understanding of financial products, concepts, and risks. Through information, instruction, and/or objective advice, they develop the skills and confidence to become more aware of financial risks and opportunities, make informed choices, know where to seek help, and take effective actions to improve their financial well-being [28].

The lack of knowledge in financial education leads many people into debt, exacerbated by the easy availability of credit for obtaining products and services [2, 22]. Francischetti et al. [20] define financial education as a process that encourages the pursuit of knowledge on how to manage money and apply resources to create wealth and financial security. The main factors affecting a person's financial competence include gender, dependents, personal and family income, and level of education [31].

Moreover, financial education also encompasses consumer pressure for sustainable practices in product production, aiming to mitigate environmental impacts [26]. In this context, studying the intersection between financial education and sustainability is essential to understand the effects on debt, savings, investments, consumption, and production [12].

In a national study, Souza and Santos [40] investigated the question "Is Financial Education a Contribution to the Economy of the 21st Century?" and concluded that financial education is crucial for the development of a more balanced society. However, they identified obstacles such as the lack of efficient public policies, inadequate teacher training, and a culture of indebtedness. The authors reaffirm that financial education can contribute to environmental preservation by deconstructing the culture of consumerism [40].

Financial education should not be seen only as a means to promote economic stability but also as a tool to build a more conscious and sustainable society. In a scenario where excessive consumption and lack of financial planning impact both individual finances and the environment, integrating sustainability concepts into financial education programmes is crucial [17]. Thus, sustainable financial education emerges as a holistic approach to addressing the socio-environmental challenges of the twenty-first century, promoting financial well-being and the preservation of natural resources for future generations [34].

In universities, financial education can play a fundamental role in developing financial skills and competences in students, promoting inclusion and ethical and sustainable principles. Empowering university students to manage their personal finances broadens their knowledge, enabling better evaluation of purchases, payment methods, types of investment, and the pursuit of financial independence. This reflects a society that values and uses money consciously [30].

3 Methodological Procedures

3.1 Profile of the University Examined

The creation of the University of Paraíba marked a significant step in higher education in Brazil, reflecting the country's commitment to expanding educational opportunities. When it transitioned to a federal institution in 1960, the establishment of the Federal University of Paraíba (UFPB) symbolised a new era of academic development and regional integration. The UFPB's multicampus structure enabled it to reach a broader student base, ensuring access to higher education across multiple cities.

The division in 2002, leading to the formation of the Federal University of Campina Grande (UFCG), was a strategic move to decentralise and specialise educational offerings. This division allowed both universities to focus on their strengths and regional needs, enhancing the quality of education and research. The UFCG's incorporation of the campuses in Campina Grande, Cajazeiras, Patos, and Sousa allowed it to cater to the specific academic and professional demands of these regions, while the UFPB continued to serve João Pessoa, Areia, and Bananeiras with its educational programmes.

The journey of the Business Administration course at UFPB highlights the challenges and triumphs of higher education in Brazil. Approved in 1962, the course faced initial delays due to infrastructure and resource constraints. Its eventual commencement in 1980 marked the beginning of a robust academic programme that adapted over the years to meet educational standards and market demands. The expansion to include a night programme in 1998 demonstrated a commitment to accessibility, accommodating students who might have work or other daytime commitments.

The official recognition of the course in 1983 by the Ministry of Education was a milestone, affirming its academic quality and relevance. The curricular reforms in 1995 and 2006 were pivotal in aligning the course with national guidelines and contemporary educational practices. These changes ensured that graduates were well-equipped with the necessary skills and knowledge to excel in both public and private sectors.

Business Administration as a field has been integral to Brazil's economic and industrial growth. The regulation of the profession in 1965 and the establishment of a minimum curriculum in 1966 underscored the importance of standardised education and professional oversight. The creation of the Regional Councils of Administration (CRAs) provided a framework for professional regulation, enhancing the credibility and integrity of the profession.

The significant increase in Business Administration courses from two before 1960 to 1805 by 2010 reflects the burgeoning demand for qualified administrators. The 2018 Higher Education Census's revelation that over 1 million students were enrolled in this field underscores its popularity and importance.

Institutions like the Getúlio Vargas Foundation (FGV) and the School of Economics, Business and Accounting at University of São Paulo (USP) have been instrumental in advancing the field of Business Administration. Their contributions to curriculum development, research, and professional training have shaped the landscape of business education in Brazil.

Since its inception in 2002, the UFCG has distinguished itself as a premier institution, particularly in Business Administration. Its focus on high-quality education and regional development has made it a cornerstone in training competent professionals who contribute significantly to the local and national economy.

3.2 Research Classification

A quantitative research approach was developed to investigate whether there is a relationship between individuals' perceptions of financial education and environmental education. Quantitative research contributes to understanding the presence and/or influence of isolated and/or associated characteristics, which can underpin actions for their prioritised addressing, as is commonly done in the foundation, planning, execution, and evaluation of public policies [27].

The study is descriptive in nature, as it seeks to describe characteristics or establish relationships between variables [21]. The population used to meet the proposed objective consisted of undergraduate students from the Business Administration course at the Federal University of Campina Grande—PB, due to their potential to become future leaders who can contribute to the integration of sustainability dimensions (environmental, economic, and social) in various contexts that impact society, such as business management and the development of public and educational policies. Thus, a convenience sample was consolidated, considering the accessibility of respondents and the period allocated for the administration of the questionnaires.

3.3 Data Collection Procedures

As defined in the first specific objective, the instrument proposed by Couto et al. [12] was used to investigate the sample's perception of the variables Financial Education (FE) and Environmental Education (DS). This instrument was structured with 45 (forty-five) questions divided into 3 (three) blocks. The first block contains 12 (twelve) questions related to the sample, namely: age, gender, marital status, course progress status, course, period, year of entry, mother's level of education, father's level of education, number of residents in the household, occupation, monthly family income, and respondent's place of residence.

The second block, related to Environmental Education, comprised 17 (seventeen) statements directed by the constructs: perception of environmental balance (S1), the influence of behaviour on natural resources (S2), understanding of biodiversity (S3), water distribution (S4) and food (S5), level of knowledge about the National Environmental System (S6, S7, and S8), deforestation (S9), importance of native populations (S10), sustainable behaviour (S11), perception of social vulnerability (S12) and sustainable economy (S13, S14, and S15), perception of citizenship in sustainability (S16, S17) [12]. As represented in Table 1, with their respective questions.

The third block, related to Financial Education, was developed through 16 (sixteen) statements directed at the constructs: understanding forms of money (FE1), the inflationary effect (FE2), means of payment (FE3), the validity of currencies (FE4), the authenticity of currencies (FE5), the National Financial System (FE6), the exchange rate (FE7 and FE8), analysis of external factors (FE9), taxes (FE10), elements of behavioural finance (FE11, FE13, and FE16), confidence and learning

Table 1 Sustainability block: constructs and relevant questions from Couto et al. [12]

S1—Perception of Environmental Balance	A healthy and balanced environment is an indispensable factor for the perpetuation of human life
S2—Influence of Behaviour on Natural Resources	The act of consuming, in general, influences the need for energy and raw materials, which can lead to the depletion of natural resources
S3—Understanding of Biodiversity	Humans need to respect and coexist within biodiversity, as they are just another form of life among many
S4—Distribution of Water	The distribution, availability, and quality of water depend on the collective effort between public, private sectors, and civil society
S5—Distribution of Food	Currently, we have the best methods, techniques, and distribution channels for food to avoid wastage
S6/S7/S8—Knowledge Level about the National Environment System	I can identify the institutions that make up the National Environment System, each of their functions, and how to contact them if necessary
	The institutions and bodies responsible for investigating and punishing environmental crimes are prepared and have the necessary resources to carry out their work as effectively as possible
	I know who to turn to and how to report a suspected or actual environmental crime
S9—Deforestation	Deforestation causes damage such as biodiversity destruction, erosion, soil impoverishment, desertification, temperature increase, floods, and silting of rivers
S10—Importance of Native Populations	Native populations contribute to environmental preservation
S11—Sustainable Behaviour	Sustainable individual habits and actions can promote large-scale collective changes
S12—Perception of Social Vulnerability	Socially vulnerable people are more exposed to environmental risks
S13/S14/S15—Sustainable Economy	It is possible to develop an economy that respects natural limits and promotes social well-being
	It is possible to generate employment and income through environmental preservation initiatives
	Sustainable institutions, both public and private, increase societal well-being
S16/S17—Perception of Citizenship in Sustainability	Environmental education and education for sustainability are essential for life
	Environmental education should be consistently addressed in education

Regarding citizenship and basic financial rights, the average for statement FE14 ("Some formal financial services are universal rights, such as accounts for receiving salaries and wages, access to debit cards, withdrawals and transfers between accounts, transaction statements, and cheque clearing, respecting the limits and rules established by current legislation") was 3.9. Statement FE15 ("I feel capable of evaluating information and making financial decisions") also had an average of 3.9.

Analysing the other questions in this block, the respondents' average level of agreement was 4.4, as represented in Table 5.

The overall average agreement of respondents concerning the statements in block 2 (Environmental Education) was 3.98, with a standard deviation of 0.39. For block 3 (Financial Education), the average was 4.06, with a standard deviation of 0.46. When performing the non-parametric Spearman correlation test to verify the relationship between the dimensions of Environmental Education (DS) and Financial Education (FE), a coefficient of 0.4 was obtained, indicating a moderate positive significance.

The average responses by dimension suggest that the students feel more knowledgeable about the Financial Education dimension, as they, on average, showed a higher degree of agreement with the statements in the financial constructs. A justification for this phenomenon could be the nature of the Business Administration course,

Table 5 Financial education

Questions	Respondents	Minimum	Maximum	Mean	Standard deviation
FE1	125	2	5	4.42	0.8158
FE2	125	1	5	4.63	0.7244
FE3	125	2	5	4.31	0.8652
FE4	125	1	5	4.16	1.0349
FE5	125	1	5	2.70	1.368
FE6	125	1	5	3.35	1.3516
FE7	125	1	5	2.98	1.1429
FE8	125	1	5	3.07	1.3148
FE9	125	1	5	3.82	1.1597
FE10	125	2	5	4.48	0.8191
FE11	125	2	5	4.69	0.6146
FE12	125	1	5	4.42	0.9181
FE13	125	2	5	4.31	0.8271
FE14	125	1	5	3.91	1.0397
FE15	125	1	5	3.90	0.9624
FE16	125	1	5	4.14	0.9479
FE17	125	2	5	4.74	0.57
FE18	125	1	5	4.60	0.7725
FE1	125	2	5	4.53	0.7136

which includes a variety of subjects focused on financial topics, whereas discussions on Sustainable Development are more limited in the curriculum (Table 6).

The first step to analysing whether there is a relationship between the dimensions of Environmental Education (DS) and Financial Education (FE) was to verify if the data followed a normal distribution. The results of the Kolmogorov–Smirnov and Shapiro–Wilk tests indicated that the data of the tested variables did not follow a normal distribution. Therefore, Spearman's correlation, a non-parametric test, was chosen to understand the behaviour of the DS and FE dimensions.

The correlation between the two dimensions was statistically significant, positive, and moderate, with a coefficient of 0.4. This suggests that the relationship did not occur by chance and that there is sufficient evidence to affirm that there is an association between these two dimensions for the Business Administration students at UFCG. The positive correlation indicates that as the values of one dimension increase, the values of the other dimension also tend to increase. In other words, students who have a higher perception or better knowledge of Financial Education also tend to have a higher perception or better understanding of Environmental Education. However, as the correlation coefficient suggests a moderate relationship, this association is not strong enough and requires further studies and other measures of analysis to be better understood.

To compare whether Business Administration students have different perceptions of the DS and FE dimensions, considering that the samples are dependent, the Wilcoxon non-parametric test was performed. The significance p-value (Sig.) of 0.03 in the Wilcoxon test indicated a significant difference between the two compared variables, meaning there is a 3% probability that the observed difference between the two samples is due to chance. Therefore, students have significantly different perceptions of these two areas. This means that the medians of the students' responses about the dimensions are different.

Additionally, the Mann–Whitney U test was performed to analyse if there is a difference in perception regarding Environmental Education and Financial Education between students in the initial and final stages of the course. For the Environmental Education dimension, the p-value was 0.136, indicating that there is not enough evidence to reject the null hypothesis. In other words, there is no statistically significant difference in the perception of environmental education between students in the initial and final stages of the Business Administration course. This may suggest that the curriculum of the Business Administration course does not consistently address the topic throughout the different periods. Discussions on sustainable development

Table 6 Descriptive statistics (sustainable development and financial education)

	N	Minimum	Maximum	Mean	Standard deviation
DS	125	1.47	4.94	3.98	0.39
FE	125	2	5	4.06	0.4
Respondents	125				

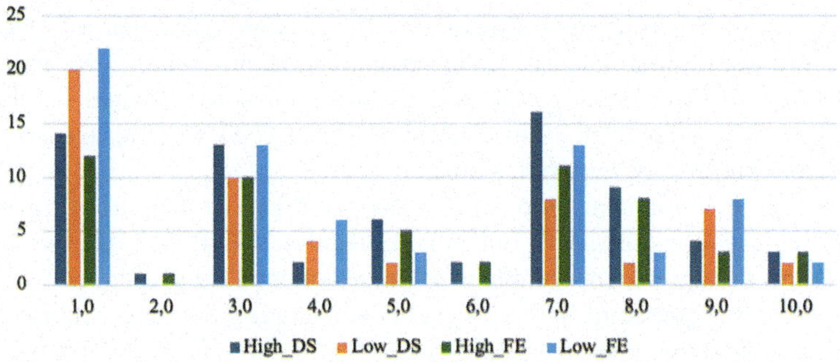

Fig. 1 Levels of sustainable development and financial education

are more extensively covered in the university's Postgraduate Programme in Administration (PPGA), especially in the concentration area of social and environmental management (Fig. 1).

The *p*-value of 0.026 for the relationship between Financial Education and Course Duration suggests that there is sufficient evidence to reject the null hypothesis. In other words, there is a statistically significant difference in the perception of Financial Education between students in the initial and final stages of the course. This difference can be attributed to a higher level of knowledge and understanding acquired throughout the course.

During the first five periods of the Business Administration course, students have the opportunity to take subjects that contribute to the development of skills and competencies related to Financial Education. Additionally, academic programmes like Pet Administration promote research projects, training, and lectures related to Financial Education, allowing students to acquire more mature or developed perceptions as they enter the final periods of the course.

These results indicate that the Business Administration course curriculum, especially in the subjects and extracurricular programmes offered in the early periods, plays a significant role in developing students' perceptions of Financial Education over time.

4.1 Research Results

The research conducted among the students of the Business Administration course at UFCG, Campina Grande campus, revealed significant insights into students' perceptions of Financial Education and Environmental Education. Here are the main findings:

- Sample Representativeness: The sample of 125 respondents, representing 31% of the population of 404 students, was solid, suggesting that the results can be generalised to the entire population of Business Administration students at UFCG.

- Perception of Environmental Education: The average agreement of respondents regarding questions related to Environmental Education was 3.98, indicating a moderate level of agreement. However, some questions related to the understanding of the national environmental system received lower averages, suggesting a possible knowledge gap in this area.
- Perception of Financial Education: The average agreement of respondents regarding questions related to Financial Education was 4.06, indicating a slightly higher level of agreement compared to Environmental Education. However, there were still some questions that received lower averages, especially those related to identifying genuine banknotes and coins.
- Correlation Between Environmental Education and Financial Education: A statistically significant, positive, and moderate correlation was found between students' perceptions of Environmental Education and Financial Education. This suggests that students who have a higher perception or better knowledge in one of these areas also tend to have a higher perception or better understanding in the other area.
- Differences in Perception Throughout the Course: There was a statistically significant difference in the perception of Financial Education between students in the initial and final stages of the course. This can be attributed to the higher level of knowledge and understanding acquired throughout the course, especially through the subjects and extracurricular programmes offered in the early periods.
- Relevance of Financial Education: The results suggest the importance of including Financial Education more comprehensively in the Business Administration course curriculum from the early stages, to better prepare students to effectively deal with financial issues in the real world.

These findings highlight the importance of comprehensively and integratively addressing both Environmental Education and Financial Education in the Business Administration course curriculum, aiming to prepare students to face the socio-environmental and financial challenges of today.

5 Conclusion and Relevant Reflections

The study aimed to investigate the relationship between individuals' perceptions of financial education and environmental education, using the instrument developed by Couto et al. [12], whose validity was confirmed by Cronbach's Alpha, ensuring the credibility of the results. The sample consisted of undergraduate Business Administration students from the Federal University of Campina Grande (UFCG)—PB, selected due to the potential of these future professionals to lead the integration of sustainability dimensions (environmental, economic, and social) in various contexts, such as business management and the formulation of public and educational policies.

The descriptive analysis of the sample profile revealed predominant characteristics, providing a context for the development of targeted educational strategies.

A relevant finding is the educational level of the students' parents, which shows a predominance of initial educational levels, potentially directly influencing the students' perception of the relevance of both financial and environmental education.

The results indicate that students' perception of financial education is more developed compared to environmental education. This discrepancy can be attributed to the greater presence of financial content in the UFCG Business Administration course curriculum, while environmental sustainability topics are more deeply covered only in the Postgraduate Programme in Business Administration. The moderate correlation between financial and environmental education suggests that improvements in financial education could benefit environmental perception, but specific efforts are needed to strengthen this integration.

The data also pointed to a significant difference in the perception of financial education between students in the initial and final stages of the course, highlighting the effectiveness of the subjects offered in the early periods. However, there was no difference in environmental perception between these groups, indicating the need for a more consistent approach throughout the curriculum.

A limitation of the research was the lack of consideration of other variables that could influence students' perceptions, such as previous experiences, exposure to media, and family environments. Additionally, the study was limited to a single institution and course, which may restrict the generalisation of the results.

Future research should explore the relationship between financial and environmental education more deeply, considering other variables and expanding the sample to include students from different courses and institutions. Furthermore, investigating the effectiveness of different teaching approaches and educational interventions in promoting economic and environmental sustainability could provide valuable insights for enhancing education in these areas.

Thus, to train more conscientious professionals, it is essential to incorporate sustainable development and environmental education topics more robustly into the university's Business Administration course curriculum, and to encourage students' participation in extracurricular projects and training related to sustainability. Educational policies that promote the integration of financial and environmental dimensions could also help to form future leaders better prepared to face the challenges of economic and environmental sustainability. This integration will benefit not only the students but also have a positive impact on organisations and society, promoting a more balanced and sustainable future.

5.1 Managerial Implications

The findings of this study have several managerial implications, especially for educational institutions and businesses. For educational institutions, the results suggest the importance of integrating financial education comprehensively into the curriculum from the early stages of the course. This integration could better prepare students to effectively handle real-world financial issues and lead to more mature perceptions of financial management. Additionally, incorporating more consistent discussions on environmental sustainability throughout the curriculum can bridge the knowledge gap observed among students. Educational programmes should also encourage student participation in extracurricular projects related to sustainability, enhancing their practical understanding and commitment to these issues.

For businesses, these findings highlight the value of recruiting graduates who possess a balanced understanding of both financial and environmental sustainability. Such individuals are likely to contribute positively to the implementation of sustainable practices within organisations. Companies should consider offering training programmes that further develop these competencies among their employees, fostering a workforce capable of addressing the complex challenges of sustainability.

5.2 *Theoretical Implications*

The study contributes to the theoretical understanding of the relationship between financial education and environmental education. The moderate positive correlation found between these dimensions suggests that enhancing financial education can positively influence environmental awareness. This finding supports the notion that financial literacy is not isolated but interconnected with other areas of knowledge, particularly environmental sustainability.

The results also indicate that curriculum design plays a crucial role in shaping students' perceptions and knowledge. The observed difference in financial education perception between initial and final-year students underscores the impact of educational content and delivery over time. This highlights the need for ongoing research into curriculum development and its effects on student outcomes across various educational dimensions.

5.3 *Sustainability Implications*

From a sustainability perspective, the study underscores the need for a more integrated approach to education that includes both financial literacy and environmental awareness. The findings suggest that students with higher financial literacy are more likely to understand and value environmental sustainability, indicating that financial education can be a gateway to broader sustainability education.

Educational institutions should aim to create programmes that simultaneously address financial and environmental topics, promoting a holistic understanding of sustainability. By doing so, they can prepare students to make informed decisions that benefit not only their personal finances but also the environment. This integrated educational approach can foster a new generation of leaders who are well-equipped to drive sustainable practices in various sectors.

5.4 Future Research Landscape

Future research should expand on the relationship between financial and environmental education by including a more diverse sample of students from different courses and institutions. This broader scope could provide a more comprehensive understanding of how these dimensions interact across various educational contexts.

Additionally, further studies should investigate the effectiveness of different teaching methods and educational interventions in enhancing both financial and environmental literacy. Experimental designs that test the impact of integrated curriculum models on student perceptions and knowledge could yield valuable insights.

Exploring the influence of external variables, such as previous experiences, media exposure, and family background, on students' perceptions of financial and environmental education would also be beneficial. This would help to identify other factors that contribute to educational outcomes and provide a more nuanced understanding of the educational process.

Overall, future research should aim to develop and test educational frameworks that promote sustainability in a comprehensive manner, ensuring that students are well-prepared to face the socio-environmental and financial challenges of the future.

References

1. Andrade K, De Oliveira JGB (2021) Educação ambiental: estratégias de conscientização ou sensibilização sobre as ações antrópicas no Brasil? In: Teixeira RLP (ed) Problemas socioambientais emergentes: contribuições teóricas e práticas. Gradus Editora, Bauru, São Paulo
2. Antunes DF (2020) As contribuições do projeto de extensão "Educação para a sustentabilidade financeira" na percepção de seus egressos. Universidade Federal do Pampa, Santana do Livramento
3. Betiol LS, Uehara THK, Laloe F, Appugliese GA, Adeodato S, Ramos L, Neto MPM (2012) Compra Sustentável: a força do consumo público e empresarial para uma economia verde e inclusiva. Programa Gestão Pública e Cidadania, FGV
4. Bourdieu P (1987) Distinction: a social critique of the judgement of taste. Harvard University Press
5. Brancalione L (2016) Educação ambiental: refletindo sobre aspectos históricos, legais e sua importância no contexto social, vol 11. Instituto de Desenvolvimento Educacional do Alto Uruguai—IDEAU
6. Brundtland GH (1987) Our common future: Report of the World Commission on Environment and Development. Oxford University Press
7. Cabral BB (2014) Educação Financeira: o primeiro passo para consumo consciente, 3rd edn. Saberes Multidisciplinares. JM Gráfica, Salvador
8. Campos CR, Coutinho CQS (2019) O juro real no contexto da educação financeira crítica. TANGRAM, Grande Dourado
9. Claudiano MS, Fiorin BH, Lopes NNL (2024) Conhecimento, Atitude e Práticas de Enfermeiros na Reanimação Cardiopulmonar: construção de um instrumento. Enferm Bras 23(1):1273–1285. https://doi.org/10.62827/eb.v23i1.pz14
10. Cohen M (2007) Crédito ao consumo, gestão financeira familiar e sustentabilidade consumo. J Int Estud Consum 31. https://doi.org/10.1111/j.1470-6431.2005.00485.x

11. Couto RF, Maracajá KFB, Batalhão A (2023) Sustainable finance: knowledge, skills and attitudes that can make a difference. Adm Ensino Pesquisa
12. Couto RF, Maracajá KFB, Machado PA (2022) Educação financeira e sustentabilidade: um framework conceitual. Adm Ensino Pesquisa 23(3):510–534. https://doi.org/10.13058/raep.2022.v23n3.2239
13. De Brasil Camargo AL (2020) Desenvolvimento sustentável: dimensões e desafios. Papirus Editora
14. Edmonson DR (2005) Likert scale: a history. Retrieved from CHARM database
15. Elkington J (1994) Towards the sustainable corporation: Win-win-win business strategies for sustainable development. California Management Review 36(2):90–100. https://doi.org/10.2307/41165746
16. Elkington J (2012) Sustentabilidade, canibais com garfo e faca. M. Books do Brasil, São Paulo
17. Ferreira Morlin M (2023) Educação Financeira: sua importância no Ensino Médio
18. Filho C, Zylbersztajn D (2003) Responsabilidade social corporativa e a criação de valor para as organizações. Rev Adm Unimep 1(1)
19. Fiorini GS (2019) Compras públicas sustentáveis: proposta metodológica e análise dos critérios socioambientais utilizados em editais de licitações de dois campi da Universidade de São Paulo
20. Francischetti CE, Camargo LSG, Santos NC (2014) Qualidade de vida, sustentabilidade e educação financeira. Rev Finanç Contab Unimep 1(1):33–47
21. Gil AC (2022) Como elaborar projetos de pesquisa, 4th edn. Atlas S/A, São Paulo
22. Gonçalves S et al (2022) A educação financeira frente ao consumo e endividamento das famílias brasileiras. Universidade Federal de Santa Catarina, Centro Sócio-Econômico
23. Khandelwal N, Darbha GK (2021) Uma década explorando MXenes como limpadores aquáticos: abrangendo uma ampla gama de contaminantes, desafios atuais e tendências futuras. Chemosphere 279. https://doi.org/10.1016/j.chemosphere.2021.130587
24. Lareau A (2011) Unequal childhoods: class, race, and family life, 2nd edn. University of California Press
25. Lotz-Sisitka H et al (2016) Co-designing research on transgressive learning in times of climate change. Curr Opin Environ Sustain 20:50–55
26. Luo W, Cheng J (2022) Transição para modelos de negócios sustentáveis para uma economia verde recuperação: papel da literacia financeira, inovação e sustentabilidade ambiental. Econ Alter Reestrut. https://doi.org/10.1007/s10644-022-09408-1
27. Mussi RFF, Mussi LMPT, Assunção ETC, Nunes CP (2019) Pesquisa Quantitativa e/ou Qualitativa: distanciamentos, aproximações e possibilidades. Rev Sustin 7
28. OECD (2005) Organização para Cooperação e Desenvolvimento Econômico (OECD). Recomendação sobre os Princípios e as Boas Práticas de Educação e Conscientização Financeira. Centro OCDE/CVM de Educação e Alfabetização Financeira para América Latina e o Caribe
29. Pignati WA, Machado JMH (2011) O agronegócio e seus impactos na saúde dos trabalhadores e da população do estado de Mato Grosso. Saúde do trabalhador na sociedade brasileira contemporânea. Editora Fiocruz, Rio de Janeiro
30. Pires BF (2008) Planejamento financeiro pessoal para estudantes universitários que estão ingressando no mercado de trabalho. Pontifícia Universidade Católica de Campinas
31. Potrich ACG, Vieira KM, Kirch G (2015) Determinantes da Alfabetização Financeira: Análise da Influência de Variáveis Socioeconômicas e Demográficas. Rev Contab Finanç 26(69)
32. Putnam RD (2015) Our kids: the American dream in crisis, reprint edn. Simon & Schuster
33. Rodrigues SCM (2019) Os recursos naturais no processo de desenvolvimento econômico capitalista: uma breve reflexão. Semiones
34. Ramos W, Barros S, Veloso L (2023) Estratégias ESG e os Objetivos de Desenvolvimento Sustentável: Framework Conceitual e de Gestão. Editora CRV
35. Romeiro AR (2012) Desenvolvimento sustentável: uma perspectiva econômico-ecológica. Estud Av 70
36. Santos JE, Sato M (2006) Universidade e ambientalismo: encontros não são despedidas. In: A contribuição da educação ambiental à esperança de Pandora, 3rd edn. Rima, São Carlos

Residential Capacity as an Indicator of Urban Sustainability for TODs: The Case Study of Tikkurila, Finland

Álvaro Corredor Ochoa

Abstract Residential density is a pivotal factor in the success of Transit-Oriented Developments (TODs). Conventionally, residential density has been measured based on the number of dwellings per unit area in "Catchment Areas" (CAs). However, this method exhibits limitations in precision, as the size of dwellings can vary significantly, impacting the potential inhabitant count. In this study, the concept of "Residential Capacity" is introduced, which assesses the number of bedrooms within a specified area, offering a more accurate estimation of the minimum and maximum population range in CAs. Using the urban environment around Tikkurila train station as a case study, residential capacity is calculated within the standard 10-min walking radius from the station but employing a realistic pedestrian catchment area measured through isochrones. The results provide a foundation for comparisons in future studies, establishing a residential capacity threshold that could serve as an indicator of urban sustainability for planners during the TOD design phase.

Keywords Sustainable urban planning · Land-use and transport planning · Sustainable transport planning · Indicators of urban sustainability · Sustainable land policy

1 Introduction

Peter Calthorpe coined the term "Transit-Oriented Development" (TOD) in 1993 in the publication "The Next American Metropolis" [8]. This urban approach combines public transportation with mixed-use residential developments, fostering the creation of walkable communities centred around transport nodes, typically train or metro stations. This model is the opposite of urban sprawl, which is characterised by its reliance on private cars, low density, and the scarcity of non-residential land uses. TODs, with their emphasis on walkability, the inclusion of green spaces, the use of

Á. C. Ochoa (✉)
Faculty of Built Environment, Unit of Architecture, Tampere University, Tampere, Finland
e-mail: alvaro.corredorochoa@tuni.fi

A. Cavalcante da Silva Batalhão and P. Singh (eds.), *Blending Approaches Towards Sustainability Focus*, Earth and Environmental Sciences Library,
https://doi.org/10.1007/978-3-031-93518-3_9

public transport, and the promotion of mixed land use, are recognised as a form of sustainable urbanism [28, 42]. They help reduce the negative impacts associated with urban sprawl, including air pollution [44], CO_2 emissions [3, 13, 16, 44], and health problems related to physical inactivity [20, 31].

Over the years, TODs have been the focus of extensive research, with over a thousand academic publications on the subject. Their implementation and study have primarily centred on regions such as North America, Europe, Australia, and Southeast Asia [27].

In TOD studies, two predominant approaches have emerged for classifying their characteristics: Robert Cervero's "3Ds" theory, later expanded to the "5Ds", and Luca Bertolini's "Node-Place" theory. The "3Ds" framework [11] highlights "Density", "Design", and "Diversity" as key attributes of TODs. This was later expanded to include destination accessibility and distance to transit, evolving into the "5Ds" [18]. In contrast, the "Node-Place" theory [6] focuses on the interaction between nodes and places as a fundamental concept.

Residential density is recognised as a fundamental and common element across all these theories for the study and development of a TOD. It is viewed as a key factor in the success of a TOD, requiring a sufficient number of users both at the transport node and in the surrounding mixed-use area [1]. However, adequate residential density alone does not ensure that residents will utilise the services and transit system of the TOD. This highlights the need for further research into residents' travel patterns, including their destinations, journey timings, preferred modes of transport, and the reasons behind their choices [35].

One of the primary objectives of a TOD is to reduce private car use and promote walking to various services within the TOD environment. This includes accessing the transit station to facilitate the use of public transport for reaching other destinations, such as workplaces or services not available within the TOD itself. As a result, a suitable mix of land uses, sufficient residential density, and high walkability around the transit node are essential factors for the effective operation and profitability of both the transport node and the associated TOD. The main indicators of residential density are typically the number of inhabitants per unit area or, more commonly in urban planning, the number of dwellings per unit area. The appropriate threshold of residential density for a TOD varies. References include Calthorpe's proposal of 10–15 housing units per acre (approximately 25–37 units per hectare) [9], and Cervero's suggestion of between 20 and 30 units per acre (50–75 units per hectare) with surface ratios of 1.0 or higher [12]. Nonetheless, the presence of a specific number of dwellings does not guarantee an exact number of residents, as these may be unoccupied, used as temporary tourist accommodations, or function as second homes. Recent urban plans, such as that of London, have introduced a second residential density indicator: the number of habitable rooms per hectare [23]. This indicator establishes different thresholds based on location within the city and accessibility to public transport (Table 1).

This approach is more realistic; however, there are factors that can cause the number of inhabitants to vary even within the same residential unit. For example, a small-sized studio without bedrooms typically houses a maximum of one or two

Table 1 Sustainable residential quality (SRQ) density matrix (habitable rooms and dwellings per hectare) [23]

Setting	Public transport accessibility level (PTAL)		
	0–1	2–3	4–6
Suburban (h/ha)	150–200	150–250	200–350
3.8–4.6 h/unit (u/ha)	35–55	35–65	45–90
3.1–3.7 h/unit (u/ha)	40–65	40–80	55–115
2.7–3.0 h/unit (u/ha)	50–75	50–95	70–130
Urban (h/ha)	150–250	200–450	200–700
3.8–4.6 h/unit (u/ha)	35–65	45–120	45–185
3.1–3.7 h/unit (u/ha)	40–80	55–145	55–225
2.7–3.0 h/unit (u/ha)	50–95	70–170	70–260
Central (h/ha)	300–650	650–1100	150–300
3.8–4.6 h/unit (u/ha)	35–80	65–170	140–290
3.1–3.7 h/unit (u/ha)	40–100	80–210	175–355
2.7–3.0 h/unit (u/ha)	50–110	100–240	215–405

people. This trend is similar in single-bedroom dwellings. However, in larger homes, the calculation becomes more complicated, as additional bedrooms can serve as studies or occasional guest rooms.

To address this issue, the use of 'residential capacity', which estimates a range of minimum and maximum habitability per dwelling, offers a more suitable and realistic approach. This method serves as a more accurate residential density indicator, linking the potential number of inhabitants with the number of dwellings.

In various countries, specific regulations apply to bedroom occupancy (Table 2) [47], allowing the residential capacity of an urban area to be calculated if the number and size of the bedrooms are known.

A key issue in TODs relates to the catchment area (CA) [4, 38, 46]. This is the residential area surrounding a TOD's transport node, acting as a hub of attraction. In it, many residents regularly travel to the transit station to use public transport or to carry out daily activities near the node. The size of the CA varies according to the mode of transport and can be defined for travel on foot, by bicycle, electric scooter, or any other sustainable means of transport. According to Calthorpe's original definition,

Table 2 Required bedroom area (m^2) for given occupancy by jurisdiction according to Viggers [47]

Allowed occupancy	New Zealand	United Kingdom	United States
1	6.00	6.50	6.50
2	10.00	10.22	9.20
3	14.00	–	13.80
4	20.00	–	18.40

the walking CA encompasses 2000 ft., equivalent to a comfortable 10-min walk. However, this time-distance relationship is not uniform, in practice, studies often define a radius of between 500 and 1000 m centred on the transit station [37, 46], with the standard distance in the US being half a mile [24].

This approach simplifies an inherently more complex process. On one hand, the use of a circle erroneously suggests the absence of obstacles and the possibility of walking in a straight line to the centre, which is seldom the case [4]. Moreover, this approach overlooks the diverse psychological and physical challenges faced by pedestrians. The psychological aspect, which is difficult to quantify, makes it impossible to accurately predict how many residents will opt to walk to the transport node and its nearby services.

There are different psychological factors influencing the choice to walk, which include weather conditions [14, 48], safety [30, 40], pedestrian infrastructure [39, 40], air pollution [32], urban land use diversity [21, 40], and residential density [41].

The factors influencing walkability range from the environment's walkability and the availability and quality of local services, to public transport features such as frequency, accessibility, service quality, and travel time to the destination [5, 15, 41, 43]. Additionally, an individual's willingness to walk depends on personal factors like the time of day, weather, and season, which impact how far one is comfortable walking [19, 48]. In contrast, physical aspects are easier to quantify, allowing for a more precise estimation of potential residents compared to the oversimplified method of calculating the population within a fixed-radius area.

Taking Calthorpe's original definition, which sets 2000 ft. as a comfortable distance for a 10-min walk, results in a walking speed of 2.3 miles/h (approximately 3.70 km/h). In contrast, using the standard half-mile CA, the speed would be 3 miles/h (about 4.80 km/h) [24].

This variation in speeds suggests the need for a more detailed analysis. It is important to consider that not all residents are capable of travelling the same distance in a given time. Also, walking speed is not constant over time [33].

Walking speed is influenced by various factors, such as age, physical condition, leg length, muscle strength, gender, and other individual-specific variables [34]. This walking speed has been extensively studied in the medical field for short distances, commonly referred to as gait speed [7]. From an urban and architectural perspective, it is primarily studied for evacuation times [45], calculating crossing times at pedestrian crossings [22], and more recently, in the context of the "15-min city" concept [10, 26, 48]. However, to the best of the current knowledge, there are no specific investigations regarding walking speed in TODs towards transportation nodes, resulting in a scarcity of research in this area.

Although there is no official definition, an isochrone map in urban planning represents the relationship between space and time for the study of a particular characteristic, commonly accessibility. This tool is widely used to calculate spatial travel time in transportation networks [33] and assess accessibility [36]. It is also employed in hydraulic management [49] or to analyse accessibility to various infrastructures, mainly in the context of healthcare facilities [50], green spaces [29], and educational institutions [10].

Although isochrone maps are a common tool in urban planning, they have not been widely used to delineate CA in TODs. Their application could be effective in linking TOD indicators with different modes of transport, including walking, cycling, electric scootering, and other sustainable means. This approach could provide valuable information on the accessibility and practicality of a TOD, thereby supporting decision-making and more detailed planning in transit-oriented development projects.

2 Data and Methodology

2.1 Study Area

In the last few decades, the vast majority of urban interventions carried out in the Helsinki metropolitan area have been jointly planned with regional mobility, primarily through commuter trains and the metro. The majority of commuter stations were already functioning in the metropolitan network as sub-centres since the last quarter of the twentieth century (Figs. 1 and 2).

These urban layouts were equipped with a certain residential density in the areas closest to the station, achieved through open residential blocks and small commercial buildings that provided basic shopping needs for the nearby population. This model of suburban urbanism, comprising semi-independent units within a municipal layout, was also replicated in the second half of the twentieth century along the main arteries connecting the city centre with surrounding municipalities. However, these examples were not planned around a railway transportation node, making the use of private cars or intercity bus lines necessary. By the late twentieth century, there was a focus on revitalizing existing railway sub-centres and creating transportation nodes in some sub-centres. Both approaches shared common denominators: the use of a shopping centre associated with the train or metro station, promoting mixed-use development in the vicinity; increasing residential density in the immediate surroundings; and a commitment to walkability to encourage frequent walking trips to the station (Fig. 3).

Vantaa, like Espoo, consists of a collection of semi-independent sub-centres that are better connected to the centre of Helsinki than to each other. Tikkurila is one of the oldest and most significant sub-centres in Vantaa [17]. Developed primarily on the west side of the railway line during the twentieth century, Tikkurila has followed regional policy guidelines, prioritizing sustainable transportation [2]. The station was one of the seven first constructed on the line connecting Helsinki to Hämeenlinna in 1862, serving as a commuter station until 1978 when a second station was built a few meters to the north [17]. Tikkurila's sub-centre has experienced significant growth mainly since the second half of the twentieth century, characterized by a mix of open housing blocks along with administrative, commercial, and basic service buildings [25]. From the early twenty-first century, Tikkurila has undergone intensive urban renewal, outlined in the 2020 General Plan, which advocates for higher building

Fig. 1 Map showing a railway network in a geographical area with labeled stations and years of establishment. Key stations include Helsinki (1862), Pasila (1890), and Tikkurila (1862). The map highlights routes with black dots representing stations, connected by green lines. Other notable stations are Kivistö (2015), Aviapolis (2016), and Lentoasema (2015). The background features a grayscale satellite view of the terrain

density with hybrid uses and emphasizes walkability around the train station. The train station itself has also undergone a significant transformation, becoming a hybrid station with a strong commercial focus, following the same pattern as other commuter train and metro stations in the Helsinki metropolitan area (Fig. 3).

2.2 *Data*

This study utilizes various quantitative GIS (Geographic Information System) data. Some of the data has been provided by the city of Vantaa. The data included in this study comprises the city map in .dwg format and diverse information for each building, including: land use category, identification code, address, building owner, building type, authorized uses, construction completion date, usable area, built area, residential area, number of floors, and the number of inhabitants. The last data point

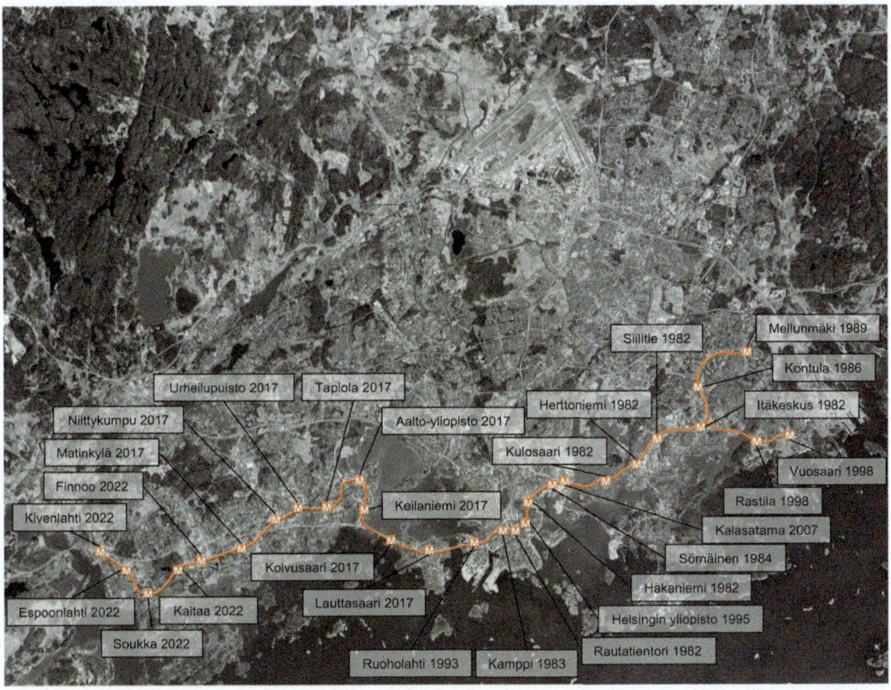

Fig. 2 Map showing a metro line in a geographical area with stations marked along the route. The line is highlighted in orange, with station names and years of establishment labeled, such as "Kamppi 1983" and "Vuosaari 1998." The map provides a visual representation of the metro network's expansion over time

is variable and connected to municipal census data, automatically updating whenever changes in residency are reported to the municipal census.

In addition, floor plans of the various buildings have been used to calculate the area of each dwelling and the number of bedrooms. The floor plans are available in various formats within the municipal GIS data.

Furthermore, Google Maps™ has been employed as a tool to obtain pedestrian distance data, both in minutes and meters.

Additionally, field observations have been conducted on foot to create a map of pedestrian zones.

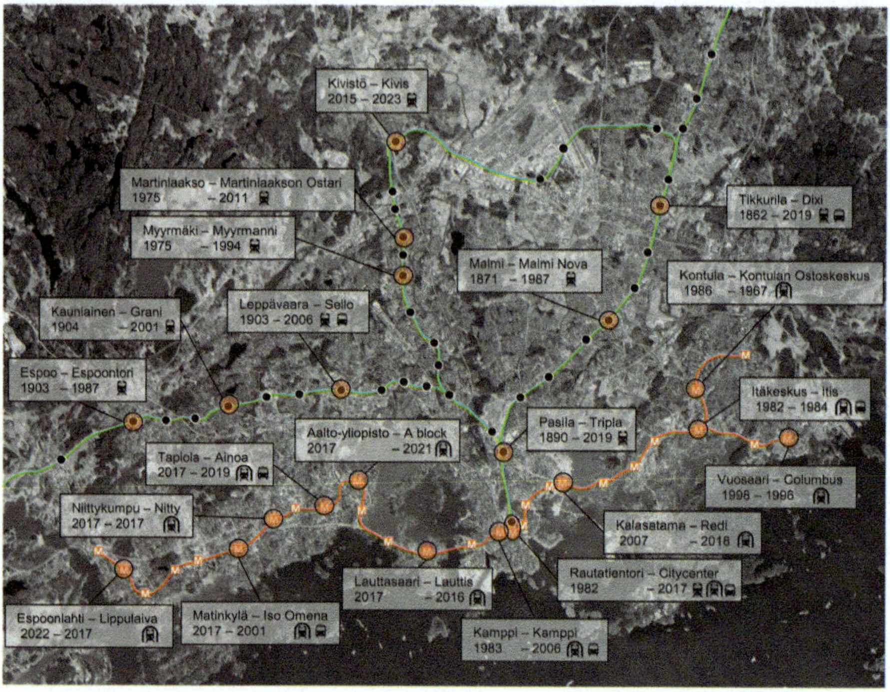

Fig. 3 Map showing a transportation network in a city, with routes marked in purple and orange lines. Key locations are labeled with names and years, such as "Kivistö 2015–2023" and "Pasila – Tripla 2019." Icons indicate transportation modes, including trains and buses. The map highlights connections between various districts, emphasizing infrastructure development over time

2.3 Calculations

2.3.1 Calculation of the Walking Catchment Area

To perform this calculation, it is necessary to obtain the pedestrian network, walking speed, and the catchment time.

The pedestrian network map was created through field observations, carefully identifying pedestrian spaces, including sidewalks, pedestrian crossings, and other pedestrian-exclusive areas. This approach provides a realistic and detailed depiction of the routes available to pedestrians (Fig. 4).

Additionally, walking speed, as previously mentioned, is influenced by factors such as age, physical condition, leg length, muscle strength, gender, and other individual-specific variables [34]. This data on walking speed is then utilised to calculate the time it takes pedestrians to reach the transport node or the area of interest.

By integrating the pedestrian network map with walking speed data, it becomes possible to define the pedestrian catchment area (CA). This area outlines the distance

Fig. 4 Pedestrian routes in the surroundings of Tikkurila station. November 2023

pedestrians can comfortably walk within a specified time from the transport node or target location. The walking CA offers important insights into the accessibility and reach of the transport node or surrounding urban areas, aiding in improved urban planning and decision-making.

According to the present understanding, there are no specific research studies on pedestrian speeds in Transit-Oriented Development (TOD) models in environments similar to this case study. Therefore, the speed study conducted by Willberg in the context of the 15-min city model [48] has been chosen to be utilized in the Helsinki metropolitan area. The concept of the 15-min city is considered complementary to the TOD model [10], presents the closest available study on walking speeds to transit stations, according to the current analysis.

The mentioned research examines the 15-min walk accessibility for different age groups, including adults and individuals aged 65 and older, in the Helsinki metropolitan area under various weather conditions. The results show a speed threshold ranging from 3.50 to 5.15 km/h. These speeds are used to calculate the isochrones in the present study.

Table 3 Minimum and maximum walking distance in 10-min-time

Time and speed	Distance (m)
10-min, speed 3.50 km/h	583
10-min, speed 5.15 km/h	858

Additionally, a 10-min catchment time has been chosen as it is commonly used in the design of TODs [24], and it is also the original time frame established by Calthorpe in the definition of TOD [8].

By applying this range of speeds, a 10-min walk translates to a range of between 583 (with a speed of 3.50 km/h) and 858 m (with a speed of 5.15 km/h) (Table 3). These isochrone distances provide insights into the walking CA around the transit station, offering valuable information for urban planning and development in the context of TODs.

2.3.2 Calculation of Residential Capacity

As a first step, it is necessary to count the number of bedrooms or habitable rooms in each dwelling. This information is not included in the GIS data, so it must be obtained by checking the building floor plans. To do this, each room with natural light that is not a bathroom, toilet, sauna, dressing room, kitchen, or living room is considered a bedroom or habitable room.

In the case of small apartments where there is only one room, it is considered a habitable space, with an occupancy of between 1 and 1.5 people.

Additionally, the following relationship is established for calculating the residential capacity, following recommendations and regulations from various countries (Table 4).

Table 4 Relationship between occupancy and bedroom area

Minimum occupancy	Maximum occupancy	Bedroom area (m^2)
1	1.5	Apartments with one room
1	2	< 9.20
2	3	9.01–13.80
3	5	13.81–18.40
4	8	> 18.40

Fig. 7 A geographical map showing an urban area with a green circle indicating Calthorpe's radius of 2000 feet (609.6 m). The map distinguishes between residential and non-residential buildings, marked in different shades of gray. The layout includes streets, blocks, and various structures within the circle, providing a detailed view of the city's infrastructure

To address these obstacles, applications like Google Maps calculate pedestrian distances from various points, resulting in an irregular surface as the CA (Fig. 12). It has been observed that the pedestrian speed in this application is unknown, although it has been noted the pedestrian zones it utilizes. These zones mostly comprise the streets where motorized vehicles pass through, which are spaces where pedestrians do not usually walk except when crossing them.

Additionally, Google does not have precise locations for building entrances, so in its calculations, it commonly uses an intermediate point of the building located directly on the street as the starting point for pedestrian calculations.

To resolve the precision issues in these results, the velocity calculations from the aforementioned Willberg study [48] has been utilized in conjunction with the actual pedestrian network (Fig. 4). This approach yields two isochrones: a minimum CA (583 m), which corresponds to a 10-min walk for individuals over 65 years old during

Fig. 8 Map showing a geographical area with a highlighted circle representing a 1/2-mile radius (804.72 m). The map distinguishes between residential and non-residential buildings using different shades. A legend in the bottom right corner indicates the color coding for building types

winter on slippery surfaces with ice, and a maximum CA (858 m), which represents a 10-min walk for adults during summer on dry surfaces (Fig. 11).

Within the 10-min CA of Google Maps, there are 295 buildings, out of which 202 are residential and 93 are non-residential (Fig. 12).

On the other hand, the maximum 10-min isochrone with a radius of 858 m includes a total of 244 buildings, out of which 168 are residential and 76 are non-residential (Fig. 13).

Finally, the minimum 10-min isochrone with a radius of 583 m includes a total of 130 buildings, out of which 84 are residential and 46 are non-residential (Fig. 14).

The results of these approximations are reflected in Table 6 and Fig. 15.

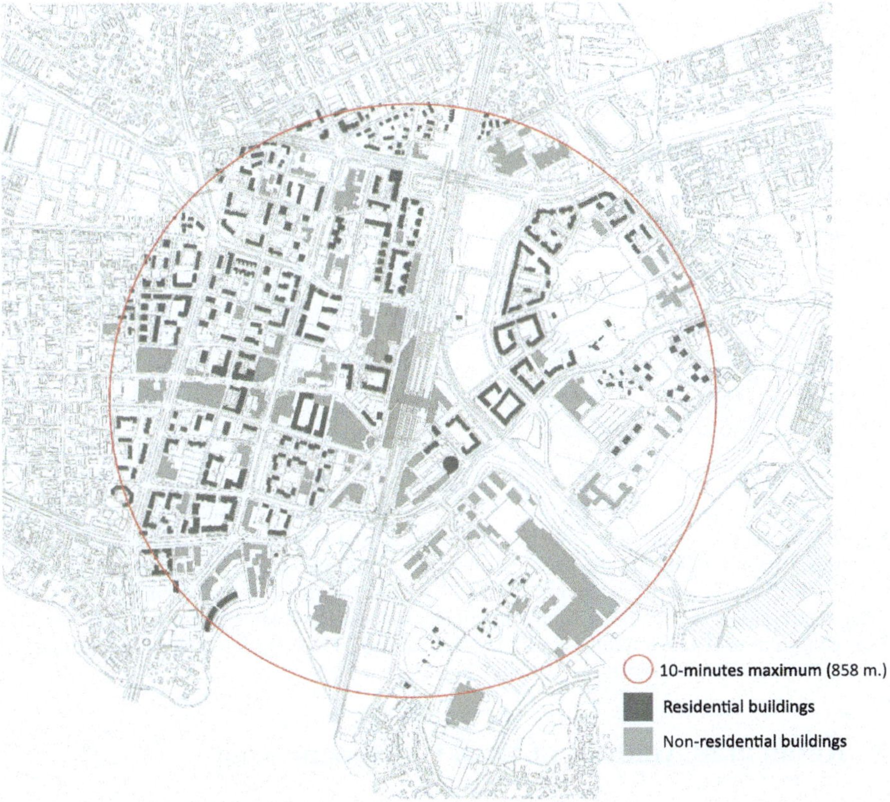

Fig. 9 Map showing a city area with a red circle indicating a 10-minute walking distance, approximately 858 m. The map distinguishes between residential and non-residential buildings using different shades of gray. Residential buildings are marked in dark gray, while non-residential buildings are in light gray. The map includes streets and other urban features, providing a detailed view of the city's layout within the specified radius

Table 5 Number of buildings within each circular catchment area (CA) approach, by use type. The CAs are defined by idealised radii: 10-minute minimum (583 m), Calthorpe's 2000 feet (609.6 m), half-mile (804.27 m), and 10-minute maximum (858 m). Distinction is made between residential and non-residential buildings

	10-min minimum 583 m	Calthorpe's 2000 ft. approach	Half-mile approach	10-min maximum 858 m
Buildings	186	209	363	432
Residential buildings	124	143	262	319
Non-residential buildings	62	66	103	113

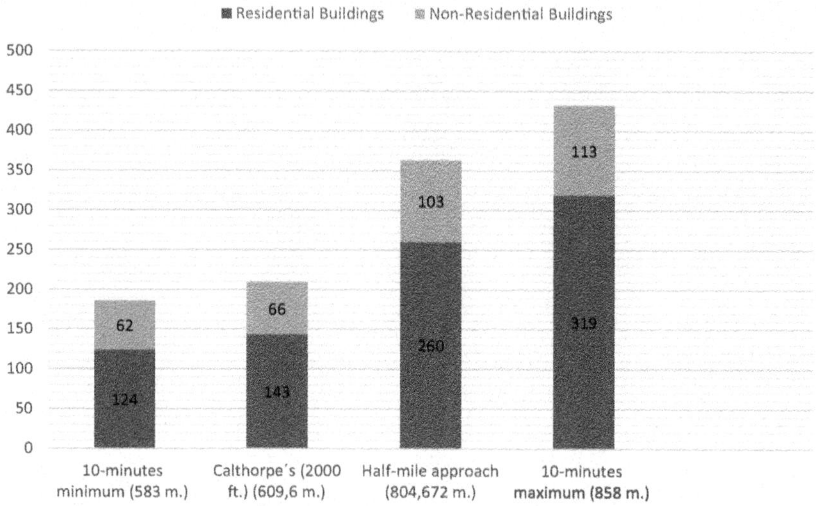

Fig. 10 Bar chart comparing the number of residential and non-residential buildings across four scenarios: "10-min minimum (583 m)," "Calthorpe's (2000 ft.) (609.6 m)," "Half-mile approach (804.672 m)," and "10-min maximum (858 m)." Each bar is divided into two segments, with darker shades representing residential buildings and lighter shades representing non-residential buildings. The values for residential buildings are 124, 143, 260, and 319, while non-residential buildings are 62, 66, 103, and 113, respectively

4 Residential Analysis of Each Catchment Area

4.1 10-min Minimum CA (583 m)

As Table 7 demonstrates, this CA covers an expanse of 60.18 ha and comprises 85 residential buildings, of which floor plans are available for 84. Among these 84, census data are only available for 70. This discrepancy arises because 14 buildings are either under construction or have only recently been completed, hence occupancy has not yet reached optimal levels and some properties may remain unsold. Within these 70 buildings, 4934 inhabitants are recorded, across 3476 dwellings, from which 4521 housing units or bedrooms are derived. The area exhibits a density of 81.99 inhabitants per hectare and a residential density of 57.76 dwellings per hectare. The residential capacity ranges from 17,178 to 4509 inhabitants, equating to a residential capacity density of between 113 and 285 inhabitants per hectare. This indicates that dwellings are occupied at below their maximum residential capacity, specifically at 71% of the maximum, yet 9% above the minimum.

Fig. 11 Map showing a city area with three outlined zones in different colors. The yellow outline represents a 583-m catchment area (CA), the blue outline indicates an 858-m CA, and the red outline shows a Google Maps 10-m CA. The map includes various buildings and streets, providing a detailed view of the urban layout

4.2 10-min Maximum CA (858 m)

This CA spans 123.64 ha and consists of 178 residential buildings, with floor plans available for 172, as outlined in Table 8. Of these 172, census data are accessible for only 144. This situation is due to some buildings being under construction or having been recently completed, resulting in suboptimal occupancy levels and the potential for some dwellings to remain on the market. Additionally, there are some single-family homes without census data. Within these 144 buildings, there are 9015 registered inhabitants and 6044 dwellings, from which 8024 housing units or bedrooms are deduced. The area has a density of approximately 73 inhabitants per hectare and a residential density of 48.88 dwellings per hectare. The residential capacity ranges from 23,888 to 8012 inhabitants, leading to a residential capacity density of between 65 and 193 inhabitants per hectare. This indicates that the dwellings are occupied at

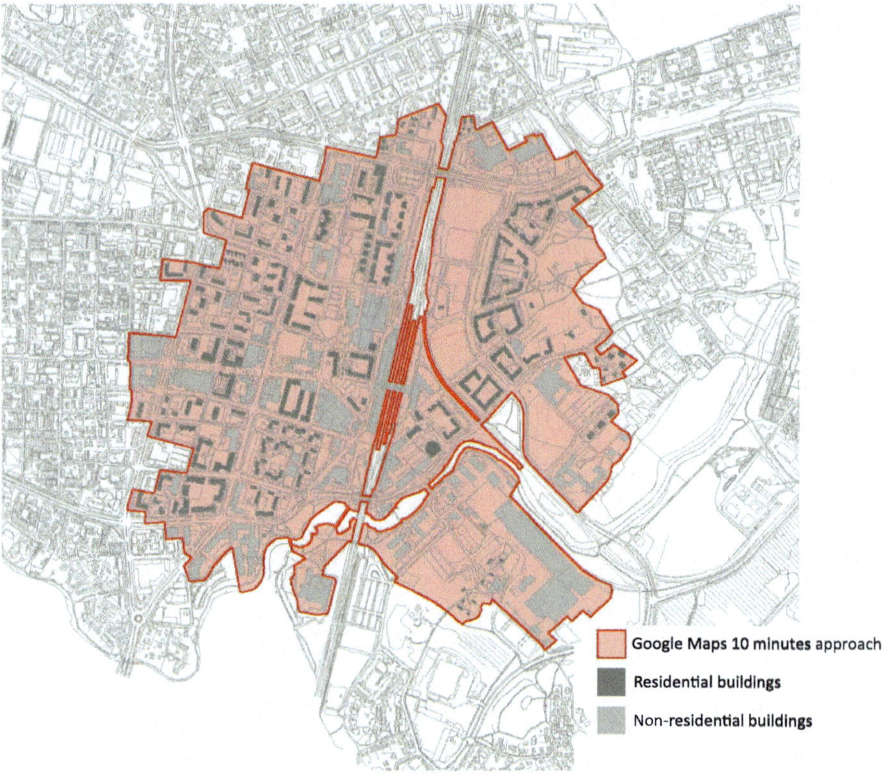

Fig. 12 Map showing a city area with highlighted zones. A pink overlay indicates a 10-min approach area according to Google Maps. The map distinguishes between residential buildings in gray and non-residential buildings in white. A legend in the bottom right corner explains the color coding. The map provides a visual representation of urban planning and accessibility

below their maximum residential capacity, specifically at 62% of the maximum, yet 13% above the minimum.

4.3 10-min CA According to Google Maps

Table 9 presents the data gathered for this CA, which spans 154.13 ha and consists of 203 residential buildings, with floor plans available for 197. Out of these 197, census data are available for only 172. This is due to the fact that some of these buildings are either under construction or have been recently completed, leading to suboptimal occupancy levels and potentially leaving some properties unsold. Additionally, there are some single-family homes without census data. In these 172 buildings, there are 9705 registered inhabitants and 6618 dwellings, from which 8758 housing units or bedrooms are derived. The residential capacity ranges between 25,257 and

Fig. 13 Map showing a geographical area with a highlighted region representing a maximum 10-min catchment area of 858 m. The map distinguishes between residential and non-residential buildings using different shades. Residential buildings are marked in dark gray, while non-residential buildings are in light gray. The highlighted area is in blue, indicating the catchment zone. The map includes detailed street layouts and building structures

8746 inhabitants. This area exhibits a density of approximately 62.97 inhabitants per hectare and a residential density of 42.94 dwellings per hectare. The residential capacity density for this CA is between 57 and 164 inhabitants per hectare. This indicates that dwellings are occupied at below their maximum residential capacity, specifically at 62% of the maximum, yet 11% above the minimum.

Fig. 14 Map showing a central area highlighted in light green, representing a 10-min catchment area of 583 m. Dark green areas indicate residential buildings, while gray areas denote non-residential buildings. The map includes a legend for reference

Table 6 Number of buildings within each network-based catchment area (CA) approach, by use type. The CAs include the 10-min isochrone from Google Maps pedestrian routing, and two GIS-based isochrones using walking speeds from Willberg et al. [48]:minimum (583 m) and maximum (858 m). Distinction is made between residential and non-residential buildings

	Google Maps 10-min CA	10-min maximum 858 m	10-min minimum 583 m
Buildings	295	250	130
Residential buildings	202	169	84
Non-residential buildings	93	82	46

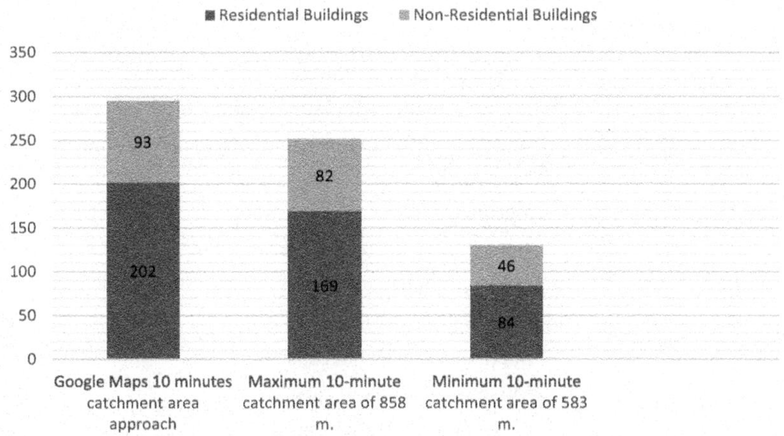

Fig. 15 Bar chart comparing the number of residential and non-residential buildings within different 10-min catchment areas. The chart includes three bars: "Google Maps 10 minutes approach" with 202 residential and 93 non-residential buildings, "Maximum 10-min catchment area of 858 m" with 169 residential and 82 non-residential buildings, and "Minimum 10-min catchment area of 583 m" with 84 residential and 46 non-residential buildings. The bars are stacked to show the total number of buildings in each category

Table 7 Residential analysis of the 583 m CA

Area (ha)	60.18
Total number of buildings	85
Buildings with floor plan information	84
Buildings with floor plan and inhabitants information	70
Inhabitants	4934
Dwellings	3476
Habitable rooms/bedrooms	4521
Density (inh/ha)	81.99
Dwelling density (dw/ha)	57.76
Maximum residential capacity	17,178
Minimum residential capacity	4509
Maximum residential capacity per ha (inh/ha)	285.44
Minimum residential capacity per ha (inh/ha)	113.48
Occupancy from maximum residential capacity (%)	71
Occupancy from minimum residential capacity (%)	− 9

Table 8 Residential analysis of the 858 m CA

Area (ha)	123.64
Total number of buildings	178
Buildings with floor plan information	172
Buildings with floor plan and inhabitants information	144
Inhabitants	9015
Dwellings	6044
Habitable rooms/bedrooms	8024
Density (inh/ha)	72.91
Dwelling density (dw/ha)	48.88
Maximum residential capacity	23,888
Minimum residential capacity	8012
Maximum residential capacity per ha (inh/ha)	193.20
Minimum residential capacity per ha (inh/ha)	64.80
Occupancy from maximum residential capacity (%)	62
Occupancy from minimum residential capacity (%)	– 13

Table 9 Residential analysis of the Google Maps CA

Area (ha)	154.13
Total number of buildings	203
Buildings with floor plan information	197
Buildings with floor plan and inhabitants information	172
Inhabitants	9705
Dwellings	6618
Habitable rooms/bedrooms	8758
Density (inh/ha)	62.97
Dwelling density (dw/ha)	42.94
Maximum residential capacity (inh)	25,257
Minimum residential capacity (inh)	8746
Maximum residential capacity per ha (inh/ha)	163.86
Minimum residential capacity per ha (inh/ha)	56.74
Occupancy from maximum residential capacity (%)	62
Occupancy from minimum residential capacity (%)	– 11

5 Analysis of the Results and Discussion

The results of the analysis of the three catchment areas (CAs), as detailed in Table 10, reveal significant differences in terms of size and other relevant parameters, with profound implications for urban sustainability regulations and transit-oriented developments (TODs).

Table 10 Summary of the analysed data according to CA

Parameters	583 m CA	858 m CA	Google Maps CA
Related walking speed (km/h)	3.50	5.15	Unknown
Increase of speed (%)	0	32	Unknown
Area (ha)	60.18	123.64	154.13
Increase of area (%)	0	51	61
Total number of residential buildings	85	178	203
Increase of total number of residential buildings (%)	0	52	58
Residential buildings with floor plan information	84	172	197
Increase of residential buildings with floor plan information (%)	0	51	57
Residential buildings with floor plan and inhabitants information	70	144	172
Increase of residential buildings with floor plan and inhabitants information (%)	0	51	59
Inhabitants	4934	9015	9705
Increase in inhabitants (%)	0	45	49
Dwellings	3476	6044	6618
Increase if dwellings (%)	0	42	47
Habitable rooms/bedrooms	4521	8024	8758
Increase in habitable rooms/bedrooms (%)	0	44	48
Density (inh/ha)	81.99	72.91	62.97
Increase in density (%)	0	− 12	− 30
Dwelling density (dw/ha)	57.76	48.88	42.94
Increase in the dwelling density (dw/ha) (%)	0	− 18	− 35
Maximum residential capacity (inh)	17,178	23,888	25,257
Increase in maximum capacity (%)	0	28	32
Maximum residential capacity per ha (inh/ha)	285.44	193.2	163.86
Increase in the maximum residential capacity per ha (inh/ha) (%)	0	− 48	− 74
Relationship between inhabitants and maximum residential capacity (%)	71	62	62
Minimum residential capacity (inh)	4509	8012	8746
Increase in minimum capacity (%)	0	44	48
Minimum residential capacity per ha (inh/ha)	74.93	64.8	56.74
Increase in the minimum residential capacity per ha (inh/ha) (%)	0	− 16	− 32
Relationship between inhabitants and minimum residential capacity (%)	− 9	− 13	− 11

The CA spanning 583 m, with a pedestrian speed of 3.50 km/h, is notably slower compared to the 858 m CA, which achieves a speed of 5.15 km/h. However, the difference in area size does not correspond proportionally; it is approximately 61% larger than the 583 m CA and 10% larger than the 858 m CA. This discovery suggests inefficient use of pedestrian routes which could be improved to facilitate access to a greater number of buildings and reduce the number of physical barriers, a crucial aspect for TODs.

The CA according to Google Maps slightly exceeds the maximum observed size, being almost 25% larger. Although precise information regarding the pedestrian speed used by Google is unavailable, it can be assumed to be similar to that of the 858 m CA, likely between 5 and 6 km/h. The lack of specific information on the pedestrian speed utilised by Google highlights an opportunity to enhance data collection and thereby the accuracy in modelling accessibility in TODs.

A notable correlation exists between the increase in the number of residential buildings and the size of the CA, showing an approximate 50% increase in the 858 m CA compared to the 583 m CA, and a 60% increase in the Google Maps CA. This demonstrates a direct relationship between the increase in CA size and the quantity of residential infrastructure. However, the rise in the number of inhabitants does not follow an identical progression, with a 45% increase observed in the 858 m CA and a 49% increase in the Google Maps CA. This discrepancy could indicate a misalignment in planning capacity to adequately meet the population demands in TOD developments.

The situation of the residents in the various catchment areas is depicted in Fig. 16, which identifies three principal areas of population concentration, designated as zones A, B, and C.

A similar pattern is observed in the number of dwellings, with a 42% increase in the 858 m CA and a 47% increase in the Google Maps CA, suggesting a direct relationship between the number of dwellings and the number of inhabitants, and likely a similar size of the dwellings. This latter point is confirmed by observing the progression in the number of housing units or bedrooms, which increase by 44% and 48% in the 858 m and Google Maps CAs, respectively. It is important to note that the size of the bedrooms or living spaces does not exceed 9.20 m^2, indicating that the residential capacity of all bedrooms is between one and two inhabitants. In this instance, the dwellings are distributed across four high-density areas (A, B, C, and D), as illustrated in Fig. 17. Zone D represents a new development, which accounts for its absence in the inhabitant count, owing to its incomplete construction.

The distribution of the number of bedrooms or habitable rooms is depicted in Fig. 18. The heat map closely resembles that of the number of dwellings, with four distinct high-density zones identified.

Density decreases as the size of the CA increases, being 12% lower in the 858 m CA and 30% lower in the Google Maps CA compared to the 583 m CA. This change in density, due to the disparity in the progression between the number of inhabitants and the area size, can be interpreted as a flaw in the planning strategy that fails to maintain optimal density levels in more extensive areas. This finding is crucial for TODs, where adequate density is essential for the model's success. A similar

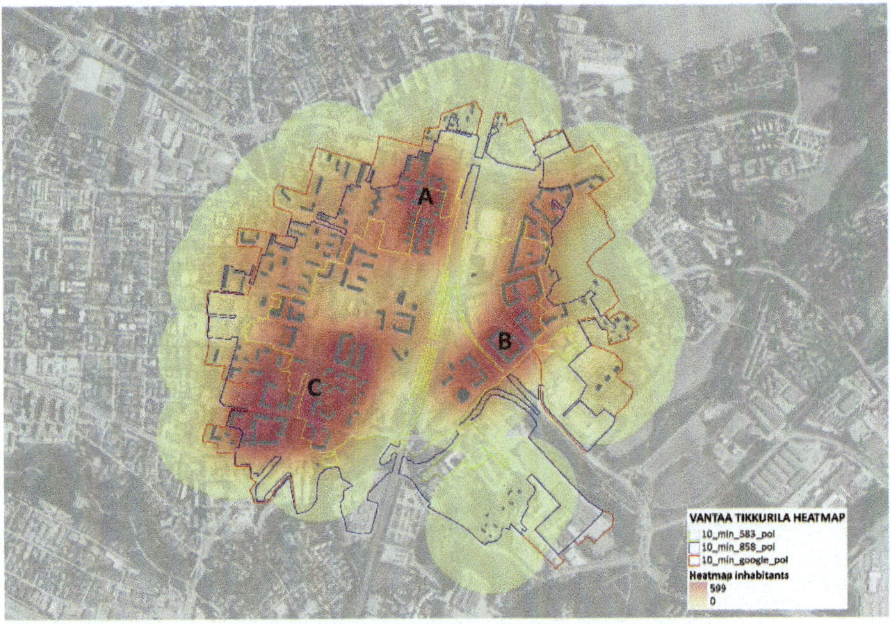

Fig. 16 Heatmap of Vantaa Tikkurila overlaid on a city map, showing varying intensity levels from red to yellow. Areas labeled A, B, and C indicate regions of higher density. A legend in the bottom right corner identifies the color codes: "10_min_583_pol," "10_min_858_pol," and "10_min_google_pol," with a scale for "Heatmap inhabitants" ranging from 0 to 599. The background displays detailed city infrastructure

explanation applies to residential density, which is reduced by 18% and 35% in the 858 m and Google Maps CAs, respectively, compared to the 583 m CA.

The maximum residential capacity shows a progressive increase across the different CAs, albeit at a slower pace compared to the increase in area or the number of dwellings and bedrooms. This suggests that dwellings may tend to be smaller as one moves away from the station. Analysing the relationship between the number of residents and the maximum residential capacity, it is observed that the smallest CA of 583 m has an occupancy of 71% of the maximum capacity, compared to 62% in the other two areas, indicating a higher maximum occupancy in the area closest to the station, an expected result given the higher observed residential density. The geographical distribution of the maximum residential capacity, presented in Fig. 19, is similar to that of dwellings and the number of rooms, though Zone B exhibits less intense shades.

Contrary to other parameters, the minimum residential capacity decreases as the size of the CA increases. In the catchment area of 858 m, the minimum capacity is 16% lower than that of 583 m, and in the Google Maps CA, it is 32% lower. This pattern suggests an increase in the minimum size of dwellings in more extensive areas, indicating a greater presence of small dwellings in the vicinity of the station. Unlike other parameters, the minimum residential capacity decreases as the size of

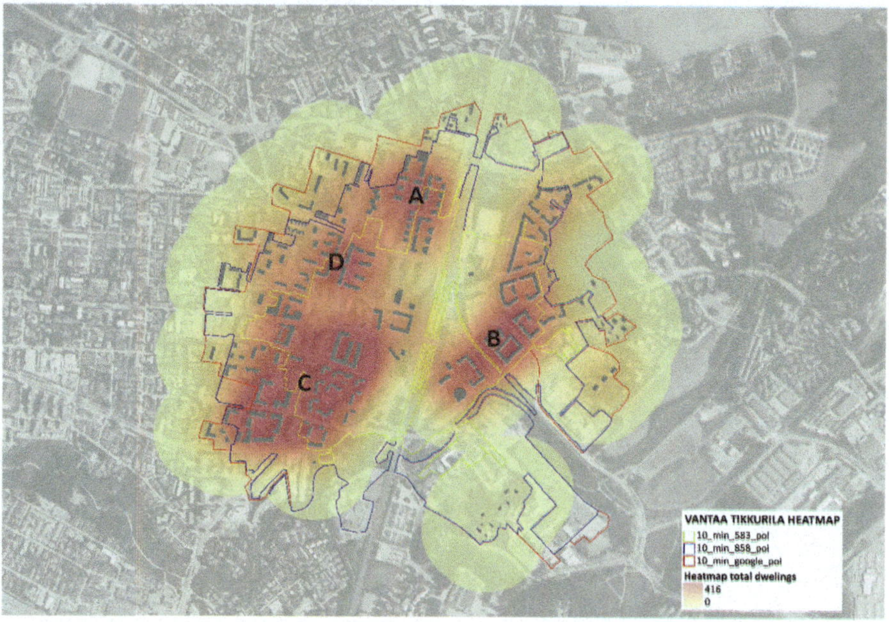

Fig. 17 Heatmap of Vantaa Tikkurila overlaid on a city map, highlighting areas A, B, C, and D. The heatmap uses a color gradient from yellow to red, indicating varying levels of intensity. A legend in the bottom right corner shows color codes for different data points: "10_min_583_pol," "10_min_858_pol," and "10_min_google_pol." The heatmap represents total dwellings, ranging from 0 to 416

the catchment area increases. In the catchment area of 858 m, the minimum capacity is 16% lower than that of 583 m, and in the Google Maps catchment area, it is 32% lower. This pattern suggests an increase in the minimum size of dwellings in larger areas, indicating a greater presence of smaller dwellings in the vicinity of the station. It could also imply lower efficiency in the use of space, thereby contradicting the principles of TODs, which aim to maximise accessibility and space efficiency. As shown in Fig. 20, the heat map of minimum residential capacity also reveals four significant areas, with Zone C exhibiting greater intensity compared to the parameter of maximum residential capacity.

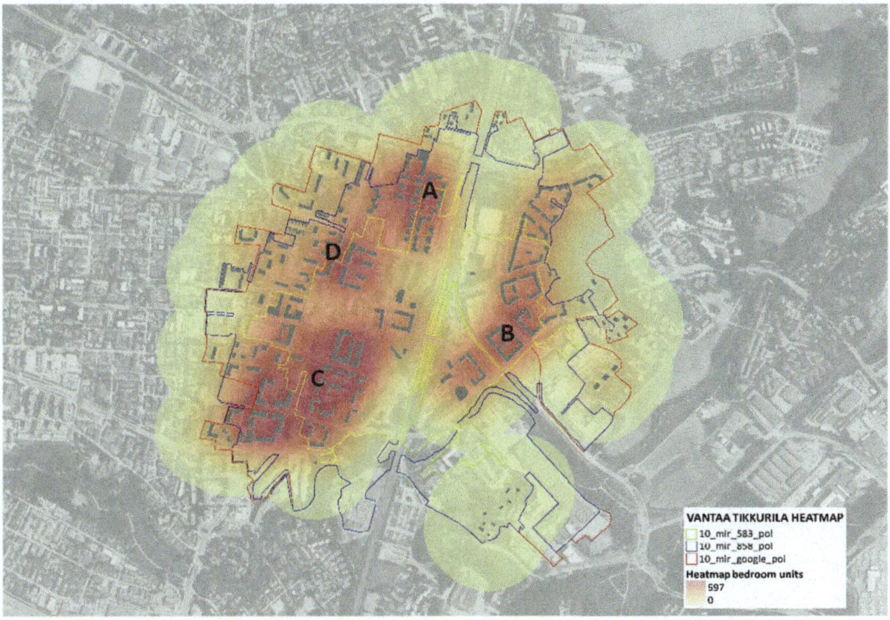

Fig. 18 Heatmap overlay on a city map of Vantaa Tikkurila, highlighting areas labeled A, B, C, and D. The heatmap uses a color gradient from yellow to red, indicating varying levels of bedroom units, with red representing higher concentrations. A legend in the bottom right corner shows color codes for different data sources: "10_mir_583_pol," "10_mir_830_pol," and "10_mir_google_pol." The heatmap scale ranges from 0 to 597 bedroom units

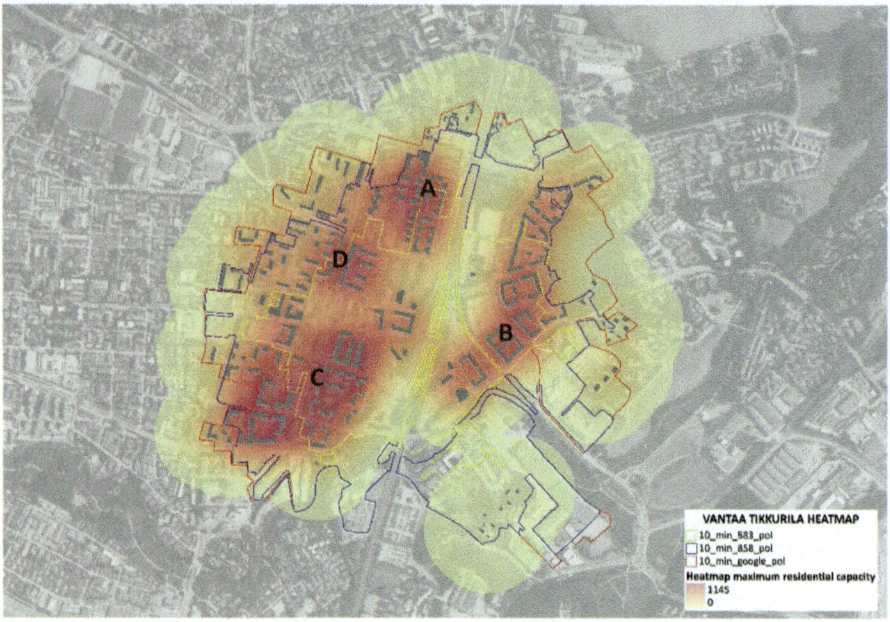

Fig. 19 A geographical map overlaying a cityscape, displaying a heatmap with varying colors from red to green, indicating different levels of data intensity. The map is divided into sections labeled A, B, C, and D. A legend in the bottom right corner explains the color coding: red for high intensity, yellow for medium, and green for low. Streets and buildings are visible beneath the overlay, providing context to the data distribution

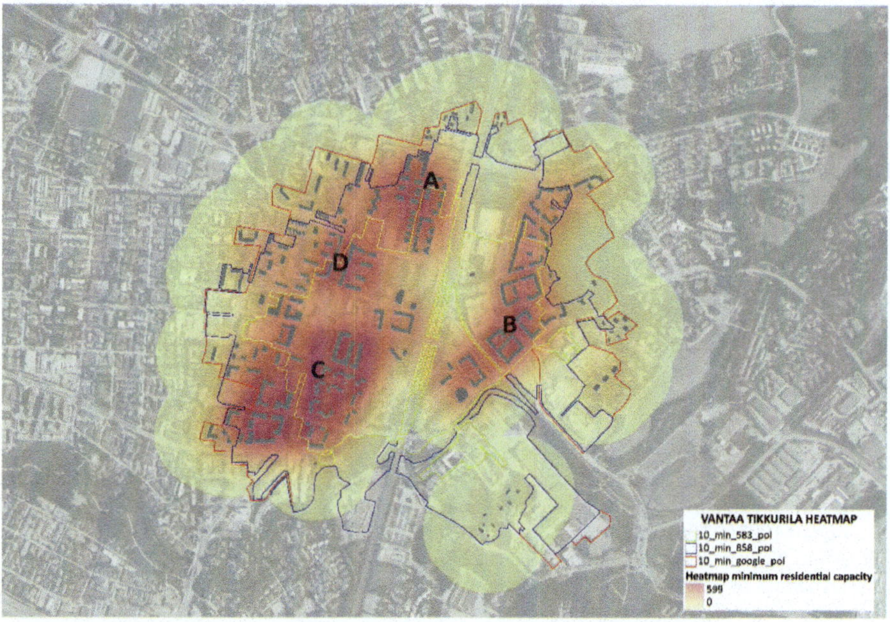

Fig. 20 Geographical map of Vantaa Tikkurila displaying a heatmap overlay. The map highlights areas with varying residential capacities, indicated by color gradients from red (high capacity) to yellow (low capacity). Key areas are labeled A, B, C, and D. A legend in the bottom right corner explains the color codes: "10_min_583_pol," "10_min_858_pol," and "10_min_google_pol," with a capacity range from 0 to 599. The background shows a detailed street and building layout

6 Conclusion

In pursuit of creating sustainable urban policy, this chapter offers an innovative approach to assessing sustainability in TODs. It focuses on one of the most important indicators of urban sustainability, namely density, adapting it specifically to TODs. Using Tikkurila as a case study, this chapter highlights the challenges associated with the classic catchment areas used in TODs, which rely on pre-established models that often deviate from reality. To overcome these challenges, it proposes the use of isochrones that define maximum and minimum areas based on variable pedestrian speeds, providing a more realistic pedestrian catchment area and offering a deeper, more specific understanding of sustainable urban planning in TODs. Additionally, it examines the issues related to the use of traditional density indicators, such as population numbers or housing density per surface area, which can vary considerably during a monitoring phase. A new indicator, residential capacity, is proposed to better reflect reality, allowing for the calculation of the maximum and minimum number of inhabitants that an urban framework can accommodate.

The approach and results presented in this chapter contribute to a better understanding of urban sustainability in TODs, paving the way for future research and establishing regulatory frameworks for evaluating TOD sustainability.

Acknowledgements I extend my gratitude to the City of Vantaa for providing data for this research. I would also like to thank Antonio Fernández Fernández and Aída Corredor Ochoa for their assistance with GIS mapping.

References

1. AlQuhtani S, Anjomani A (2021) Do rail transit stations affect the population density changes around them? The case of Dallas-Fort Worth metropolitan area. Sustainability 13:3355. https://doi.org/10.3390/su13063355
2. Albacete X, Olaru D, Paül V, Biermann S (2017) Measuring the accessibility of public transport: a critical comparison between methods in Helsinki. Appl Spat Anal 10:161–188. https://doi.org/10.1007/s12061-015-9177-8
3. Ali L, Nawaz A, Iqbal S, Aamir Basheer M, Hameed J, Albasher G, Shah SAR, Bai Y (2021) Dynamics of transit oriented development, role of greenhouse gases and urban environment: a study for management and policy. Sustainability 13:2536. https://doi.org/10.3390/su13052536
4. Badland H, White M, MacAulay G, Eagleson S, Mavoa S, Pettit C, Giles-Corti B (2013) Using simple agent-based modeling to inform and enhance neighborhood walkability. Int J Health Geogr 12:58. https://doi.org/10.1186/1476-072X-12-58
5. Berawi MA, Saroji G, Iskandar FA, Ibrahim BE, Miraj P, Sari M (2020) Optimizing land use allocation of transit-oriented development (TOD) to generate maximum ridership. Sustainability 12:3798. https://doi.org/10.3390/su12093798
6. Bertolini L (1999) Spatial development patterns and public transport: the application of an analytical model in the Netherlands. Plan Pract Res 14:199–210. https://doi.org/10.1080/02697459915724
7. Bohannon RW (1997) Comfortable and maximum walking speed of adults aged 20–79 years: reference values and determinants. Age Ageing 26:15–19. https://doi.org/10.1093/ageing/26.1.15
8. Calthorpe P (1993) The next American metropolis: ecology, community, and the American dream. Princeton Architectural Press, New York
9. Calthorpe P, Fulton WB (2001) The regional city: planning for the end of sprawl. Island Press, Washington, DC
10. Caselli B, Carra M, Rossetti S, Zazzi M (2022) Exploring the 15-minute neighbourhoods. An evaluation based on the walkability performance to public facilities. Transp Res Procedia 60:346–353. https://doi.org/10.1016/j.trpro.2021.12.045
11. Cervero R, Kockelman K (1997) Travel demand and the 3Ds: density, diversity, and design. Transp Res Part D Transp Environ 2:199–219. https://doi.org/10.1016/S1361-9209(97)00009-6
12. Cervero R, Murphy S, Ferrell C, Goguts N, Tsai Y-H, Arrington GP, Boroski J, Smith-Heimer J, Golem R, Peninger P, Nakajima E, Chui E, Dunphy R, Myers M, McKay S, Witenstein N (2004) Transit-oriented development in the United States: experiences, challenges, and prospects. Transportation Research Board, Washington, DC. https://doi.org/10.17226/23360
13. Chatman DG (2013) Does TOD need the T?: On the importance of factors other than rail access. J Am Plann Assoc 79:17–31. https://doi.org/10.1080/01944363.2013.791008
14. Delclòs-Alió X, Marquet O, Vich G, Schipperijn J, Zhang K, Maciejewska M, Miralles-Guasch C (2019) Temperature and rain moderate the effect of neighborhood walkability on walking time for seniors in Barcelona. IJERPH 17:14. https://doi.org/10.3390/ijerph17010014

15. Dell'Olio L, Ibeas A, Dominguez A, Gonzalez F (2012) Passenger preference analysis: light rail transit or bus versus car. Transport 27:276–285. https://doi.org/10.3846/16484142.2012.719839
16. Dou Y, Luo X, Dong L, Wu C, Liang H, Ren J (2016) An empirical study on transit-oriented low-carbon urban land use planning: Exploratory Spatial Data Analysis (ESDA) on Shanghai, China. Habitat Int 53:379–389. https://doi.org/10.1016/j.habitatint.2015.12.005
17. Eskola A (2012) Seuraavana Tikkurila! = Nästa Dickursby = The next stop Tikkurila. Vantaan kaupunginmuseo
18. Ewing R, Cervero R (2001) Travel and the built environment: a synthesis. Transp Res Rec 1780:87–114. https://doi.org/10.3141/1780-10
19. Fossum M, Ryeng EO (2021) The walking speed of pedestrians on various pavement surface conditions during winter. Transp Res Part D Transp Environ 97:102934. https://doi.org/10.1016/j.trd.2021.102934
20. Frank LD, Fox EH, Ulmer JM, Chapman JE, Braun LM (2022) Quantifying the health benefits of transit-oriented development: creation and application of the San Diego Public Health Assessment Model (SD-PHAM). Transp Policy 115:14–26. https://doi.org/10.1016/j.tranpol.2021.10.005
21. Frank LD, Sallis JF, Conway TL, Chapman JE, Saelens BE, Bachman W (2006) Many pathways from land use to health: associations between neighborhood walkability and active transportation, body mass index, and air quality. J Am Plann Assoc 72:75–87. https://doi.org/10.1080/01944360608976725
22. Gates TJ, Noyce DA, Bill AR, Van Ee N (2006) Recommended walking speeds for timing of pedestrian clearance intervals based on characteristics of the pedestrian population. Transp Res Rec 1982(1):38–47. https://doi-org.libproxy.tuni.fi/10.1177/0361198106198200106 (Original work published 2006)
23. Greater London Authority (2015) The London plan. The spatial development strategy for London consolidated with alterations since 2011. https://www.london.gov.uk/sites/default/files/the_london_plan_2016_jan_2017_fix.pdf
24. Guerra E, Cervero R, Tischler D (2012) Half-mile circle: does it best represent transit station catchments? Transp Res Rec 2276:101–109. https://doi.org/10.3141/2276-12
25. Haavisto FH (2021) Tikkurilan kaavarunkoalueen modernin rakennuskannan inventointi 2020. https://www.vantaa.fi/sites/default/files/matti/923129-913611-062700%20Tikkurilan%20modernin%20rakennuskannan%20inventointi%202020%2C%2019.5.2021.pdf
26. Hosford K, Beairsto J, Winters M (2022) Is the 15-minute city within reach? Evaluating walking and cycling accessibility to grocery stores in Vancouver. Transp Res Interdiscip Perspect 14:100602. https://doi.org/10.1016/j.trip.2022.100602
27. Ibraeva A, Correia GHDA, Silva C, Antunes AP (2020) Transit-oriented development: a review of research achievements and challenges. Transp Res Part A Policy Pract 132:110–130. https://doi.org/10.1016/j.tra.2019.10.018
28. Knowles RD, Ferbrache F, Nikitas A (2020) Transport's historical, contemporary and future role in shaping urban development: re-evaluating transit oriented development. Cities 99:102607. https://doi.org/10.1016/j.cities.2020.102607
29. Kolcsár RA, Szilassi P (2018) Assessing accessibility of urban green spaces based on isochrone maps and street resolution population data through the example of Zalaegerszeg, Hungary. Carpath J Earth Environ Sci 13:31–36. https://doi.org/10.26471/cjees/2018/013/003
30. Lamour Q, Morelli AM, Marins KRDC (2019) Improving walkability in a TOD context: spatial strategies that enhance walking in the Belém neighbourhood, in São Paulo, Brazil. Case Stud Transp Policy 7:280–292. https://doi.org/10.1016/j.cstp.2019.03.005
31. Langlois M, Wasfi RA, Ross NA, El-Geneidy AM (2016) Can transit-oriented developments help achieve the recommended weekly level of physical activity? J Transp Health 3:181–190. https://doi.org/10.1016/j.jth.2016.02.006
32. Marshall JD, Brauer M, Frank LD (2009) Healthy neighborhoods: walkability and air pollution. Environ Health Perspect 117:1752–1759. https://doi.org/10.1289/ehp.0900595

33. Miller HJ, Bridwell SA (2009) A field-based theory for time geography. Ann Assoc Am Geogr 99:49–75. https://doi.org/10.1080/00045600802471049
34. Mohamed O, Appling H (2020) Clinical assessment of gait. In: Orthotics and prosthetics in rehabilitation, 4th edn. Elsevier, pp 102–143. https://doi.org/10.1016/B978-0-323-60913-5.00005-2
35. Olaru D, Moncrieff S, McCarney G, Sun Y, Reed T, Pattison C, Smith B, Biermann S (2019) Place vs. node transit: planning policies revisited. Sustainability 11:477. https://doi.org/10.3390/su11020477
36. O'Sullivan D, Morrison A, Shearer J (2000) Using desktop GIS for the investigation of accessibility by public transport: an isochrone approach. Int J Geogr Inf Sci 14:85–104. https://doi.org/10.1080/136588100240976
37. Pongprasert P, Kubota H (2019) TOD residents' attitudes toward walking to transit station: a case study of transit-oriented developments (TODs) in Bangkok, Thailand. J Mod Transport 27:39–51. https://doi.org/10.1007/s40534-018-0170-1
38. Sander HA, Ghosh D, Van Riper D, Manson SM (2010) How do you measure distance in spatial models? An example using open-space valuation. Environ Plann B Plann Des 37:874–894. https://doi.org/10.1068/b35126
39. Schlossberg M, Brown N (2004) Comparing transit-oriented development sites by walkability indicators. Transp Res Rec 1887:34–42. https://doi.org/10.3141/1887-05
40. Shigematsu R, Sallis JF, Conway TL, Saelens BE, Frank LD, Cain KL, Chapman JE, King AC (2009) Age differences in the relation of perceived neighborhood environment to walking. Med Sci Sports Exerc 41:314–321. https://doi.org/10.1249/MSS.0b013e318185496c
41. Sohn K, Shim H (2010) Factors generating boardings at metro stations in the Seoul metropolitan area. Cities 27:358–368. https://doi.org/10.1016/j.cities.2010.05.001
42. Su S, Zhang H, Wang M, Weng M, Kang M (2021) Transit-oriented development (TOD) typologies around metro station areas in urban China: a comparative analysis of five typical megacities for planning implications. J Transp Geogr 90:102939. https://doi.org/10.1016/j.jtrangeo.2020.102939
43. Su S, Zhao C, Zhou H, Li B, Kang M (2022) Unraveling the relative contribution of TOD structural factors to metro ridership: a novel localized modeling approach with implications on spatial planning. J Transp Geogr 100:103308. https://doi.org/10.1016/j.jtrangeo.2022.103308
44. Suzuki H, Murakami J, Hong Y-H, Tamayose B (2015) Financing transit-oriented development with land values: adapting land value capture in developing countries. The World Bank. https://doi.org/10.1596/978-1-4648-0149-5
45. Thompson PA, Marchant EW (1995) A computer model for the evacuation of large building populations. Fire Saf J 24:131–148. https://doi.org/10.1016/0379-7112(95)00019-P
46. Tong X, Wang Y, Chan EHW, Zhou Q (2018) Correlation between transit-oriented development (TOD), land use catchment areas, and local environmental transformation. Sustainability 10:4622. https://doi.org/10.3390/su10124622
47. Viggers H (2016) Occupancy inefficiency of larger detached houses. In: Miller W, Susilawati C, Manley K (eds) HealthyHousing2016: proceedings of the 7th international conference on energy and environment of residential buildings, Nov 2016. Queensland University of Technology, Australia, Brisbane. https://doi.org/10.4225/50/58107299bcb89
48. Willberg E, Fink C, Toivonen T (2023) The 15-minute city for all?—Measuring individual and temporal variations in walking accessibility. J Transp Geogr 106:103521. https://doi.org/10.1016/j.jtrangeo.2022.103521
49. Yang G, Bowling LC, Cherkauer KA, Pijanowski BC (2011) The impact of urban development on hydrologic regime from catchment to basin scales. Landsc Urban Plan 103:237–247. https://doi.org/10.1016/j.landurbplan.2011.08.003
50. Zhang Y, Li W, Deng H, Li Y (2020) Evaluation of public transport-based accessibility to health facilities considering spatial heterogeneity. J Adv Transp 2020:1–10. https://doi.org/10.1155/2020/7645153

SDG, Smart City and Pandemic Issue: Evaluation in Campina Grande, Brazil

Stephanie Freire Brito, Maria de Fátima Martins, and André Cavalcante da Silva Batalhão

Abstract Cities are the localities where the majority of the world's population is concentrated and are also responsible for activities that directly impact global sustainability. Their ability to innovate and adapt to new contexts can suggest new concepts to define them, such as 'Smart Cities'. Observing these modifications and adaptations to new contexts, this study is dedicated to analysing the congruence of the characteristics of the city of Campina Grande, Brazil through the sectors of a Smart City. For this purpose, an evaluation model for Smart Cities, aligned with the SDGs and the COVID-19 pandemic experienced from 2020 onwards in the locality, is applied. The main results demonstrate alignment with important sectors for Smart Cities such as ICTs and Education, Culture, Innovation and Science. However, it also shows misalignment with sectors such as energy, natural environment, and built environment. The evaluation enables the development of actions and policies to improve city services and provide greater benefits to the population residing there.

Keywords Smart City · SDGs · COVID-19 · Indicators · Campina Grande

1 Introduction

The concepts of sustainable development and sustainability arise from the perception of a deteriorated environmental scenario [6]. Management based on sustainable development necessitates the use of instruments that indicate its practice in the

S. F. Brito · M. de Fátima Martins (✉)
Graduate Program in Administration at UFCG, Campina Grande, Campina Grande, Paraíba, Brazil
e-mail: atima.martins@uaac.ufcg.edu.br

A. C. da Silva Batalhão
Graduate Programme in Environmental Sciences, Federal University of Goiás, Goiânia, Goiás, Brazil

Department of Applied Management Sciences, Minas Gerais State University, Passos, Minas Gerais, Brazil

A. Cavalcante da Silva Batalhão and P. Singh (eds.), *Blending Approaches Towards Sustainability Focus*, Earth and Environmental Sciences Library,
https://doi.org/10.1007/978-3-031-93518-3_10

process of diagnosis and alignment of decision-making. For this purpose, sustainability indicators emerge, aiming to systematically provide the profile of the analysed object, incorporating social, economic, political, institutional, environmental, demographic, cultural aspects, etc.

Narrowing the debate on sustainable development management to the locality level, it is possible to observe that the decision-making process is preceded by data collection subjected to an analysis process, enabling an understanding of the reality directed at municipalities, highlighting emerging aspects and future trends.

Cities renew themselves over time. The integration of technological factors into the traditional structures of cities leads to the emergence of Smart Cities, a model incorporating a higher level of information and communication technology into the urban centre and the population's daily life [5], contributing to the achievement of the Sustainable Development Goals (SDGs) [11] and addressing the challenges posed by the COVID-19 pandemic.

Therefore, the purpose of this study is to operationalise and evaluate the level of sustainability of the city of Campina Grande in Paraíba, regarding its performance concerning the characteristics of a Smart City, the fulfilment of the SDGs, and the response to the COVID-19 pandemic. The instrument used for this purpose will be a set of indicators constructed based on the ISO 37-122:2019 standards, SDG targets, and explicit demands from the literature, allowing for verification through a comparison between the effectiveness of actions and the levels deemed ideal from the perspective of the proposed and validated model, considering local particularities.

The process of collecting, processing, and analysing the indicators will follow the methodology guidelines, allowing for the data to be made available through indices and resulting in a classification of municipalities regarding their respective levels of sustainability.

2 Methodological Procedures

This study is characterised as quantitative, descriptive, and exploratory, aiming to understand social phenomena through the location, evaluation, and synthesis of data and information over a specific period. Its objective is to measure the level of sustainability of the city of Campina Grande in Paraíba regarding its performance concerning the characteristics of a Smart City, the fulfilment of the SDGs, and the response to the COVID-19 pandemic.

The research will be conducted based on secondary data collected under the methodology focused on the SDGs for cities with Smart City characteristics and pandemic resilience, developed to measure the sustainability of localities from this perspective. It is a quantitative study, as it will utilise quantitative data collected by research institutes and obtained from public agencies and institutions.

Due to its objectives, the research can also be classified as descriptive, as there is a need to describe the diagnoses obtained through a set of sustainability indicators. Finally, it is characterised as exploratory due to its updating nature, using recent and

Table 1 Sequence of activities

No.	Sequence of activities
1	Define the set of variables
2	Collect secondary data and information through research institutes and governmental bodies
3	Identify the unit of measurement for each variable
4	Transform the variables into indices within their respective dimensions
5	Identify the negative or positive relationship of each variable in the context of sustainable development
6	Classify the variables according to their relationship

previously unknown data. Regarding the methodological procedures, the research will adopt guidelines consisting of the following sequence of activities (Table 1).

The obtained data were transformed into indices that indicated the city's level of intelligence, pandemic resilience, and compliance with the SDGs through the analysis of indicators developed in matrices for each of the 11 dimensions defined according to the dimensions of Smart Cities as designated by Huovila et al. [10], and Abu-Rayash and Dincer [1].

The results are classified by two types of relationships, either positive or negative. In this specific application, each indicator within each of the matrices and themes is evaluated positively or negatively. In the case of a positive result (positive relationship), it indicates more alignment with the characteristics of Smart Cities, alignment with the SDGs, and better pandemic resilience. In the case of a negative result (negative relationship), it indicates less alignment with the characteristics of Smart Cities, alignment with the SDGs, and better pandemic resilience, highlighting critical points for improvement.

2.1 Evaluation Object—Campina Grande, Paraíba

The city of Campina Grande is the second-largest city in the state of Paraíba and one of the main cities in the interior of the Northeast. It has a significant urbanisation level, with 94.96% of its population in urban areas. According to Aesa [2], its HDI-M-2000 is 0.592, and the municipality stands out as number 3 in the state of Paraíba. In 2015, it had a population of 405,072 inhabitants, representing 10% of the state's population, with a GDP corresponding to 14% of the state's GDP.

Alongside the Metropolitan Region of João Pessoa, the Metropolitan Region of Campina Grande exhibits dynamics that could be considered metropolitan. The two represent the most important population arrangements in the state and are configured as Regional Capitals I and II, respectively. It is noteworthy that, according to the IBGE, both regions show dependent relationships with Recife, the largest centre and Metropolis in this network.

The city has a significant history of technological, knowledge, and innovation aspects, being the first city in the North and Northeast of Brazil to have a computer installed in 1967. In the 1970s and 1980s, there was a strong expansion of UFPB activities in Campina Grande, leading to the city's emergence as a relevant university hub in the interior of the country. Also, in the 1980s, the Paraíba Technological Park was founded, and in 1988, the Technology-Based Business Incubation Program (ITCG) was created by Fapesq.

Nowadays, the main institutions that comprise the CeT base in Campina Grande linked to the innovation economy, specifically the software sector, include the Federal University of Campina Grande (UFCG), especially the Centre for Science and Technology (CCT), which includes the Department of Systems and Computing (DSC) and the Department of Electrical Engineering (DEE), both accredited by the Committee on Information Technology (CATI) to develop projects and supported by the Information Technology Law.

Additionally, the city's name has been used in media vehicles as being a Smart City. Examples include the Smart Cities Index by the Easypark group, which ranks Smart Cities globally, evaluating 500 cities from various countries, and the Urban Systems/Sator Connected Smart Cities Ranking, which evaluates cities in Brazil. Campina Grande is highlighted as the second-largest city in the state of Paraíba, with around 400,000 inhabitants and a per capita GDP of R$ 19,696.95, and it is among the 100 best Brazilian Smart Cities listed in the 2020 Ranking [13].

Regarding the COVID-19 pandemic, no performance evaluation in scientific sources could be found at the time of this research's application, except for the statistical data presented by research bodies, without structured performance analyses.

2.2 *Selection of Indicators for Analysis in Campina Grande, Paraíba*

Sustainability indicators are necessary to measure, evaluate, and facilitate progress towards a broad range of social, environmental, economic, and institutional goals. They also help identify problems and set sustainability development goals. Besides supporting decision-making, they assist managers in fund allocation, optimal use of natural resources, comparison between processes and situations, identifying trends, providing warning information, and anticipating future conditions.

For the scope of this research, the chosen indicators aim to measure and evaluate the sustainability level of the city of Campina Grande in Paraíba concerning its performance regarding the characteristics of a Smart City, compliance with the SDGs, and response to the COVID-19 pandemic. The choice of indicators is based on data availability from research institutes, public bodies, and institutions.

Out of a set of 187 indicators, 65 will be used that fit the current reality and have available data (Table 2).

Table 2 Evaluation indicators for Smart Cities, SDGs, and the pandemic

Theme	Indicators
Natural environment	Existence of specific legislation to address environmental issues; population access to real-time public alert systems regarding air and water quality (%); territory designated for conservation purposes
Built environment	Public recreation services that can be reserved online; public schools with internet access; existence of renewable energy sources; existence of smart buildings; recently renovated and installed public lighting (%)
Water and waste	Smart structure for monitoring water health problems; existence of water supply systems; existence of water treatment systems; number of selective solid waste collection programs; existence of a solid waste management plan according to PNRS terms
Transport	Tickets or electronic cards for public transport use; municipal buses with GPS and real-time location information; public transport accessibility (%); public transport services covered by a unified payment system (%); smart traffic lights (%); public transport routes with internet connectivity (%)
Energy	Energy produced from wastewater treatment, solid waste, and other liquid waste treatments and residual heat resources (%); decentralised electricity production systems (%); existence of energy efficiency policies for local economic activities; alternative system in case of power interruptions
Economy	Institutional purchase of local producers; workforce employed in education and research and development sectors (%); workforce employed in the information and communication technology (ICT) sector (%); existence of financing and credit services for small entrepreneurs; existence of public consortium, partnership agreements, private sector or community support for urban development, employment, education, health, culture, tourism, and environment
Education, culture, innovation and science	Existence of a basic education plan for digital inclusion; number of institutions promoting technological development education; research and technology investment in clean, renewable energy and energy efficiency; existence of entrepreneurship support and training programs; municipal budget invested in science and technology; existence of technological incubators; existence of a technological park
Health, well-being, and safety	Homeless population (%); hospital beds per thousand inhabitants (%); number of public basic units; existence of municipal health plans for physical and psychological health for women; city area covered by digital surveillance cameras; existence of cooperation policies between municipalities for health demands
Governance and citizen engagement	Women in command of municipal government departments (%); women elected as councillors (%); black individuals elected (%); number of municipal conferences with citizen participation; public employees investigated for corruption (%); existence of a performance indicator monitoring system by municipal management

(continued)

Table 2 (continued)

Theme	Indicators
ICT	Operational transport control centre; water quality monitoring technologies; city's water distribution network monitored by a smart water system (%); existence of automation and electric grid control systems; existence of smart energy measurement systems accessible to consumers; city area mapped by streets in real-time interactive systems (%)
Pandemic resilience	Vaccination planning; pandemic response rate (tests/day); existence of exclusive infrastructure for COVID-19 patients; number of ICU beds for COVID-19; existence of an educational plan for adaptation to digital continuity of the academic year; existence of support programs for economically affected activities by the pandemic; actions to protect poor and vulnerable people; existence of COVID-19 contagion monitoring systems; availability of public data on resources directed to combat COVID-19; number of corruption investigations related to COVID-19 resources; public participation in decisions on COVID-19 combat programs and pandemic resilience

2.3 Data Collection

The secondary data for each of the indicators were catalogued from databases that provide information on Brazilian cities, including IBGE; Municipal Government of Campina Grande, Paraíba; Datasus; Anatel; Regional Electoral Tribunal; National Treasury Secretariat; Ministry of Cities; Ministry of Culture.

3 Results and Discussion

To achieve the research objectives, the analysis in the city of Campina Grande utilised 10 dimensions of Smart Cities as designated by Huovila et al. [10], and the dimension of "Pandemic Resilience" by Abu-Rayash and Dincer [1]. The indicators were chosen through a review of ISO 37-122:2019 indicator standards, SDG targets, and literature demands, categorised within the dimensions.

The 11 management sectors of a Smart City were analysed, with the following matrices: (1) Natural environment; (2) Built environment; (3) Water and waste; (4) Transport; (5) Energy; (6) Economy; (7) Education, culture, innovation and science; (8) Security; (9) Governance and citizen engagement; (10) ICT; (11) Pandemic resilience.

Matrix 1, Natural environment, evaluates the dimension of natural resources and attention to their preservation. In a Smart City, technology emerges as a means to monitor and manage these resources to increase sustainability levels. The results for this sector are provided in Table 3.

is an evident gap in the measurement of workforce numbers in important sectors for the Smart City economy, which are education, research and development, and ICT skills [4].

The seventh matrix, "Education, Culture, Innovation, and Science," refers to a sector of the Smart City. The indicator matrix aids in monitoring investment and management in the education and human development sector, which is an important requirement for the economic and social development of the locality, encouraging people to interact with the city's smart resources and contributing to the SDG goals in a global purpose. The results for this sector are provided in Table 9.

The sector demonstrates a positive relationship in most indicators. Ahvenniemi et al. [3] highlight that this sector of the Smart City should aim to capitalise on the education system's policy, creating more opportunities for students and teachers using ICT tools. The evaluation shows that the city performs well in this sector, making it a strong point for highlighting as a Smart City.

The eighth matrix refers to the "Health, Well-being, and Safety" sector. The operationalisation of these indicators generates information about the health, well-being, and safety aspects of citizens in the urban space concerning its structure and the control of the main problems generated in this area. The results for this sector are provided in Table 10.

According to Neirotti et al. [12], the health, well-being, and safety sector of a Smart City relies on ICT resources to optimise processes and promote more efficiency

Table 9 Indicators for the education, culture, innovation and science sector

SDG	Sector: Education, culture, innovation and science			
	Indicators	Data	Source	Relation
	Existence of a basic education plan for digital inclusion	Yes	SEDUC	Positive
	Number of institutions that promote teaching technological development	24 (twenty four)	SEDUC	Positive
	Investment in research and technology for clean, renewable energy and energy efficiency	Information unavailable	SEFIN/SEDUC	Negative
	Existence of support and training programs for entrepreneurship	4 (four)	SECTI	Positive
	Municipal budget invested in science and technology	Information unavailable	SEFIN/SECTI	Negative
	Existence of technological incubators	Yes	SECTI	Positive
	Existence of a technology park	Yes	SECTI	Positive

Table 10 Indicators for the health, well-being and safety sector

SDG	Sector: Health, well-being and safety			
	Indicators	Data	Source	Relation
	Homeless population (%)	Information unavailable	IBGE/ Health's Secretary	Negative
	Hospital beds per thousand inhabitants (%)	0.046 (zero point zero four six)	DATASUS	Positive
	Number of public basic units	82 (eighty-two)	DATASUS	Positive
	Existence of municipal physical and psychological health plans for women	Yes	Social assistance secretary	Positive
	Area of the city covered by digital surveillance cameras	Information unavailable	Municipal guard	Negative
	Existence of cooperation policies between municipalities to meet health demands	Yes	Health's Secretary	Positive

in activities. The matrix of this sector demonstrates positive relationships in most of its indicators; however, the variable that mentions the technological resource of the matrix lacks information for evaluation, suggesting a point of attention for the competent managers.

The ninth matrix, "Governance and Citizen Engagement," aims to evaluate popular participation in the governance aspects of a Smart City, as well as various SDG-related goals such as the empowerment of women and the reduction of racial inequalities, along with participatory governance. The results for this sector are provided in Table 11.

The matrix demonstrates negative relationships in half of the indicators, highlighting problems in the governance structure regarding participatory governance and anti-corruption requirements, which are characteristic of a Smart City and in alignment with the SDG goals [1]. The observed representativeness with the participation of women and black individuals in high-ranking positions in the city's public power is important for formulating projects and policies directed at these groups, cooperating with the SDG goals.

The tenth matrix aims to measure the Information and Communication Technology (ICT) sector of the Smart City. ICTs operate in the city in an integrated manner with its resources and services, offering greater utilisation of activities and efficiency of services. It is a strong characteristic of Smart Cities [7]. The results for this sector are provided in Table 12.

According to Vanolo [14], in a Smart City, traditional networks and services are more flexible, efficient, and sustainable through the use of ICTs. The matrix indicators

Table 11 Indicators for the governance and citizen engagement sector

SDG	Sector: Governance and citizen engagement			
	Indicators	Data	Source	Relation
	Women in charge of municipal government departments (%)	22.5%	PMCG	Positive
	Women elected councillors (%)	30.5%	PMCG	Positive
	Black people elected (%)	17%	PMCG	Positive
	Number of municipal conferences with population participation	Information unavailable	PMCG	Negative
	Public officials investigated for corruption (%)	Information unavailable	PMCG	Negative
	Existence of a system for monitoring performance indicators by municipal management	No	PMCG	Negative

Table 12 Indicators for the ICT sector

SDG	Sector: ICT			
	Indicators	Data	Source	Relation
	Transport operational control center	Yes	STTP	Positive
	Water quality monitoring technologies	Yes	CAGEPA	Positive
	Smart water system monitoring the water distribution network (%)	No	CAGEPA	Negative
	Existence of electrical network automation and control systems	Yes	Energisa	Positive
	Existence of an intelligent energy metering system for consumer access	Yes	Energisa	Positive
	City area mapped by streets in real-time interactive systems (%)	Yes	STTP	Positive

suggest that the use of ICTs in the city has been positive regarding the supply of essential services such as energy, water, and mobility, with the only shortcoming being the integration of information and communication technologies in water distribution monitoring. This is a point of attention for managers in developing work plans.

The eleventh matrix refers to "Pandemic Resilience," with indicators assessing the city's capacity to adapt through necessary resources and activities to manage the critical outbreak, such as the COVID-19 pandemic. The results for this sector are provided in Table 13.

Abu-Rayash and Dincer [1] state that a Smart City demonstrates resilience in the face of pandemics and national disasters involving loss of life, with an effective response rate. Moreover, it uses all available resources to minimise loss of life and support the well-being of affected citizens in economic, social, and mental aspects.

The results of this matrix were somewhat balanced in terms of their relationship classification. From these indicators, it is possible to ascertain that the city managed to articulate actions that supported the pandemic's demand concerning specific care

Table 13 Indicators for the pandemic resilience sector

SDG	Sector: Pandemic resilience			
	Indicators	Data	Source	Relation
	Vaccination planning	Yes	PMCG/ Health's Secretary	Positive
	Pandemic response rate (tests/day)	Information unavailable	PMCG/ Health's Secretary	Negative
	Existence of exclusive infrastructure to care for those infected by COVID	Yes	PMCG/ Health's Secretary	Positive
	Number of ICU beds for COVID	131 (one hundred thirty-one)	PMCG/ Health's Secretary	Positive
	Existence of an educational adaptation plan to continue the school year via digital means	Yes	SEDUC	Positive
	Existence of support programs for economic activities harmed by the pandemic	Yes	SECTI/ SEDE	Positive
	Protection actions for poor and vulnerable people	No	SEMAS	Negative
	Existence of COVID contagion monitoring systems	No	Health's Secretary	Negative
	Provision of public data on resources aimed at combating COVID	Yes	PMCG/ Health's Secretary	Positive
	Number of cases investigated for corruption relating to resources to combat COVID	1 (one)	Public Ministry	Negative
	Popular participation in decisions about programs to combat COVID and pandemic resilience	Information unavailable	PMCG/ Health's Secretary	Negative

structures and socio-economic problems, albeit without records of popular participation in these decisions. On the other hand, besides the unavailability of important data such as the pandemic response rate through testing, there are records of corruption investigations involving the city's handling during the health crisis.

Overall, the evaluation shows a misalignment of Campina Grande with the management sectors of Natural Environment, Built Environment, and Energy, with predominantly negative relationships in the indicators of these sectors. On the other hand, there is greater conformity of the city with the sectors of Water and Waste; Education, Culture, Innovation, and Science; and the ICT sector. Regarding the Transport, Economy, Health, Well-being and Safety, Governance and Citizen Engagement, and Pandemic Resilience sectors, the city shows a certain balance, with both positive and negative relationships being quite similar.

4 Conclusion

The present research is developed with the aim of evaluating sustainability in the city of Campina Grande through a model that considers the evaluation dimensions of a Smart City and its connection with the collaboration towards achieving the SDG targets, in addition to considering the pandemic resilience of city management during the critical period experienced with the advancement of COVID-19 worldwide from the year 2020.

The results obtained from the analysis conducted with the city of Campina Grande show that some progress has been made for the city to be characterised as a Smart City. The evaluation demonstrates the city's strengths and weaknesses regarding its "Smartmentalisation," a term used by Vanolo [14] to refer to the process of adapting elements that make cities smarter and leaders in social, technological, and environmental development.

In this context, and in view of the results obtained through this research, greater criticality is indicated in relation to the Natural Environment, Built Environment, and Energy; medium criticality in the sectors of Transport, Economy, Health, Well-being and Safety; Governance and Citizen Engagement, and Pandemic Resilience; and low criticality in the sectors of Water and Waste, Education, Culture, Innovation and Science, and ICT. This finding is useful for developing growth strategies to achieve higher levels of "Smartisation" of the city, as well as mutual collaboration with the SDGs' targets and addressing critical moments, as the city demonstrated when dealing with the COVID-19 pandemic.

References

1. Abu-Rayash A, Dincer I (2020) Analysis of the electricity demand trends amidst the COVID-19 coronavirus pandemic. Energy Res Soc Sci 68:101682
2. Aesa (2016) Divisão Hidrográfica e Hidrogeológica Associadas às Regiões de Desenvolvimento. Governo do Estado da Paraíba. Available at: http://www.aesa.pb.gov.br/aesa-website/wp-content/uploads/2016/11/PE_06.pdf. Accessed 20 June 2024
3. Ahvenniemi H, Huovila A, Pinto-Seppä I, Airaksinen M (2017) What are the differences between sustainable and smart cities? Cities 60:234–245
4. Albino V, Berardi U, Dangelico RM (2015) Smart cities: definitions, dimensions, performance, and initiatives. J Urban Technol 22(1):3–21
5. Angelidou M (2014) Smart city policies: a spatial approach. Cities 41:S3–S11
6. Barbieri JC, Silva D (2011) Desenvolvimento sustentável e educação ambiental: uma trajetória comum com muitos desafios. Rev Adm Mackenzie São Paulo SP 12(3):51–82
7. Bibri SE, Krogstie J (2017) Smart sustainable cities of the future: an extensive interdisciplinary literature review. Sustain Cities Soc 31:183–212
8. Caragliu A, Del Bo C, Nijkamp P (2011) Smart cities in Europe. J Urban Technol 18(2):65–82
9. Giffinger R, Pichler-Milanović N (2007) Smart cities: ranking of European medium-sized cities. Centre of Regional Science, Vienna University of Technology
10. Huovila A, Bosch P, Airaksinen M (2019) Comparative analysis of standardized indicators for smart sustainable cities: what indicators and standards to use and when? Cities 89:141–153
11. Meschede C (2019) Information dissemination related to the sustainable development goals on German local governmental websites. Aslib J Inf Manag. https://doi.org/10.1108/ajim-08-2018-0195
12. Neirotti P, De Marco A, Cagliano AC, Mangano G, Scorrano F (2014) Current trends in smart city initiatives: some stylised facts. Cities 38:25–36
13. Urban Systems (2020) Ranking connected smart cities. Retrieved from https://d335luupugsy2.cloudfront.net/cms/files/48668/1600973008Ranking_CSC_2020.pdf. Accessed 7 June 2021
14. Vanolo A (2014) Smartmentality: the smart city as disciplinary strategy. Urban Stud 51(5):883–898

Smart Cities, Sustainable Development Goals, and COVID-19: Can Paths Be Interwoven?

Stephanie Freire Brito, Maria de Fátima Martins, and André Cavalcante da Silva Batalhão

Abstract The largest portion of the world population is concentrated in cities, requiring management models to make them inclusive, safe, resilient, and sustainable, as suggested by SDG 11. Ensuring these predicates for humanity has become even more urgent in the face of the COVID-19 pandemic. Observing this scenario, the question arises: Do Smart Cities cooperate with the goals of the SDGs? If so, does this cooperation apply to the pandemic scenario? Bibliometrics assists in mapping the field of study in this sense, showing links between the themes in scientific works, and content analysis examines contexts used to make this link in theoretical and empirical works. The analysis reveals the ascendancy of an emerging theme, which deals with the connection of Smart Cities with the SDGs in four main areas of analysis: Intelligent Processes and Learning; Social Innovation; Sustainability; and Waste Management. The interpretations in each of the contexts are varied. The study looks at these contexts and reveals opportunities for future research.

Keywords Smart Cities · SDG · COVID-19 · Bibliometrics · Content analysis

1 Introduction

The evolution of the environmental debate has resulted in alliances involving various global communities. Following the launch of the Global Compact and the creation of the Millennium Development Goals (MDGs) in 2000, a union of 193 United Nations

S. F. Brito · M. de Fátima Martins (✉)
Graduate Program in Administration at UFCG, Campina Grande, Campina Grande, Paraíba, Brazil
e-mail: atima.martins@uaac.ufcg.edu.br

A. C. da Silva Batalhão
Graduate Programme in Environmental Sciences, Federal University of Goiás, Goiânia, Goiás, Brazil

Department of Applied Management Sciences, Minas Gerais State University, Passos, Minas Gerais, Brazil

A. Cavalcante da Silva Batalhão and P. Singh (eds.), *Blending Approaches Towards Sustainability Focus*, Earth and Environmental Sciences Library,
https://doi.org/10.1007/978-3-031-93518-3_11

member countries in 2015 approved an action plan for the next 15 years, called the 2030 Agenda. The plan is composed of 17 Sustainable Development Goals (SDGs) and 169 targets, which serve as guidelines for driving actions and management policies aimed at sustainable growth and greater social justice worldwide.

Among the 17 goals, one exclusively focuses on urban areas, SDG 11, titled "Sustainable Cities and Communities," aiming to make cities more inclusive, safe, resilient, and sustainable [55]. This focus coincides with the characteristics of Smart Cities, which integrate information and communication technologies (ICT) into traditional infrastructures, fostering economic growth, enabling quality of life, and benefiting the environment [14, 18].

The Smart Cities approach highlights the importance of investing in technologies to support the sustainable development of a locality [15, 24], identifying its relevant role in implementing policies to promote sustainable urban development. The discussion on the monitoring structures in Smart Cities for achieving climate goals is demonstrated by Rasca and Waeben [59] and emphasized by Ismagilova et al. [32] as a potential force for delivering many of the SDGs.

Furthermore, humanity is currently witnessing a sanitary and psychosocial catastrophe [19] caused by the COVID-19 pandemic, which spreads due to the physical contact between people, making population concentration an urgent issue [25]. In other words, it is an urban problem that can compromise the achievement of the SDGs, impacting various sectors of a Smart City [1].

Discussions involving the relationship between Smart Cities as a pathway to achieving the SDGs [32, 59] and the clear impact of the pandemic on their sector performance [1] demand the analysis of this thematic scope to highlight pathways for decision-making and stimulate new research that develops it conceptually and empirically. Not having identified other research in this sense, this chapter explores and maps studies related to Smart Cities as a tool for achieving the SDGs, including the pandemic, contributing to identifying research on the topic and demonstrating its contributions to this area of study.

The bibliometric analysis enables this research to find answers to the following problems: (a) What are the main themes, authors, co-authorship networks, journals, universities, and countries in this field of research? (b) How are publications on Smart Cities for achieving the SDGs classified?

This study advances research on Smart Cities, SDGs, and the pandemic by defining and integrating the field of study among the themes, contributing to understanding its current state by identifying the main publications, as well as the co-authorship networks that reveal research links. Furthermore, the study uses qualitative content analysis to identify implicit and underlying perspectives between the two themes. The aim of this chapter is to highlight Smart Cities' practices as tools for achieving the SDGs, including the pandemic period, through a systematic or integrative review.

2 Theoretical Framework

2.1 Smart Cities and SDGs: Theoretical Path

The participation of Smart Cities in the scientific field is somewhat recent [33]. Its concept was considered imprecise by Hollands [30], confusing by Caragliu et al. [18], and in development by Serey et al. [65]. However, the integration of ICTs into traditional city infrastructure is a predominant factor in literary analyses [2, 15, 45, 52, 67].

Based on the literature review conducted by Neirotti et al. [52], researchers Huovila et al. [31] designate 11 predominant sectors in Smart City: Natural Environment, Built Environment; Water and Waste; Transport; Energy; Economy; Education, Culture, Innovation, and Science; Health, Well-being, and Safety; Governance and Citizen Engagement; ICT. In the pandemic context, Abu-Rayash and Dincer [1] add Pandemic Resilience as an evaluation and investment sector.

Harvey [29] points out that cities are a reflection of the capitalist society we live in, indicating the basis of its context of evolution in the mechanics of extraction, production, and consumption, which has historically resulted in social, environmental, and economic problems [61]. However, the technological aggregation of Smart Cities is considered a potential model to optimise this scenario and provide higher levels of sustainability [15, 16, 22, 52].

The perception of a deteriorated environmental scenario [12] results in global agreements aimed at repairing and preventing the intensification of these damages [23]. Thus, sustainable development emerges as a response to a purely capitalist development model [44], culminating, among other initiatives, in the creation of the 17 goals and 169 targets of the SDGs [55], which direct the creation of policies and research around various sustainability-related themes [37].

One of these themes includes the context of cities, which according to Merida and Lacerda [46], demands sustainable solutions to face urban planning and management challenges. Greater urgency is attributed to this context due to the accelerated urbanisation process [27], making cities the future environment for humanity, as there is an expectation that 70% of the world's population will reside in urban centres by 2050 [74].

This thesis is reinforced by the SDG 11 initiative, which aims to "Make cities and human settlements inclusive, safe, resilient, and sustainable" [55], highlighting its influence on accelerating prosperity and well-being [5], and its relevant role in building a new development model [42]. Thus, the technological presence characteristic of Smart Cities is an effective tool for sustainability, positively impacting social, environmental, and economic dimensions [39, 46, 57].

The concept of Smart City is considered an attractive policy for cities, and SDG 11 specifically supports its popularity [31], so much so that the New Urban Agenda (NUA) reinforces cities' commitment to sustainability, aligning their priorities with the SDGs [34, 51]. The sectors of Smart Cities dialogue with the SDGs [32], and their successful implementation will contribute to achieving a large part of their targets [3, 5, 22].

The concept of Smart City is primarily, but not exclusively, linked to SDG 11. Several studies aim to show examples of how smart city initiatives enabled by ICT can contribute to achieving, in the context of sustainable city development, all 17 SDGs, complementing and interdependent with the global scenario [5, 31, 36, 59, 69].

2.2 *The Smart Cities Scenario in Controlling the COVID-19 Pandemic*

In the context of the pandemic, an important observation to be made regarding SDG 11 is its point 11.1, which states "ensuring access for all to basic services" and point 11.5, which indicates that "by 2030, significantly reduce the number of deaths and the number of people affected by disasters, and substantially decrease the direct economic losses caused by them" [55]. SDG 11 does not specifically refer to the term pandemic; however, according to de Castilho Sá et al. [19], the COVID-19 pandemic experienced in 2020 is characterised as a sanitary and psychosocial catastrophe, as it imposes vital damages on almost the entire global population, with catastrophe being the closest term to name it.

Abu-Rayash and Dincer [1] affirm that this virus affected all aspects of life in cities, including the economy, social life, politics, the environment, the technology sector, and the health sector. Innovation, knowledge, and resilience are intrinsic aspects of Smart Cities [21, 49, 78], which operate with the common ultimate goal of providing quality of life to citizens [46].

It is observed that cities with smart aspects, Smart Cities, have made more progress in controlling and achieving better results in addressing the COVID-19 pandemic in 2020 [1, 41]. The need for rapid action impacts public administrations [77] and requires resilience from policies and the population, as it profoundly changes their way of life. Therefore, resilience is an expected skill at this time [17], especially from the health system during the pandemic, which is an essential factor for maintaining social and economic activities [43, 68].

For example, the municipal managers of Wuhan, amidst the epidemic outbreak, took actions by installing intelligent infrastructures to enhance their security and information services, converting it into a Smart City. The new mechanisms helped contain and monitor the disease outbreak in the city with the implementation of Big Data ecosystems, alerting about people and vehicle flows in real-time [41].

The multiplicity of elements in a city interfered with by the effects of the pandemic [1] requires better understanding from managers. In this case, Smart Cities are committed to data sharing, making managers' work produce better results [4]. The theoretical framework demonstrates the relevance of the SDGs for constructing sustainable growth, including in the urban context, which contributes to more resilient environments, providing the community with more robust basic structures.

3 Methodological Procedures

To analyse the relationship between Smart Cities, SDGs, and the implications during the pandemic, a systematic or integrative review was used, approached by Souza et al. [70] as a methodological approach that allows the inclusion of diverse methods, aiming to identify, evaluate, and interpret all relevant research on a specific topic. This technique enables the synthesis of produced knowledge and provides support for the scientific field, assisting in formulating questions and proposals for future research [66].

To achieve this purpose, this integrative review used bibliometrics [54] and the assistance of the VOSviewer software, Bardin's content analysis [13] through Iramuteq, and manual reading of the main articles indicated by the software. Bibliometrics is a measurement technique that evaluates scientific production activities, allowing a contextual understanding of these productions and their authors [9]. The aim of mapping the field of study of Smart Cities, SDGs during the pandemic, and suggesting research agendas in this scope is achieved through this approach.

According to Vanti [76], some primary objectives of bibliometrics are: (a) identifying trends and growth in scientific knowledge production in an area; (b) predicting publication trends; (c) identifying individual productivity of authors, institutions, and countries; (d) analysing citation and co-citation processes; (e) understanding the possibilities for developing new research.

To meet these objectives and categories of analysis, mapping and clustering techniques were used, obtained through the VOSviewer software [75], which allowed the identification of the most relevant researchers, journals, and research institutions within the network, in addition to generating maps using co-citation, co-authorship, and country co-authorship analysis. The data analysis, network, and map creation were generated with the help of VOSviewer version 1.6.7, conducting the analyses: (1) co-authorship networks; (2) citation networks; and (3) co-citation networks. The objective of this methodology is to show the structural and dynamic aspects of the three themes together, understanding academic production in terms of authors cited together through co-citation analysis, with the structure of citations in terms of common patterns [73].

The unit of analysis consists of scientific articles published between 2015 and 2020 in the main collection of the Web of Science (WoS) database. The search was conducted on January 31, 2021. The articles were filtered by the terms "Smart Cit*" and "Sustainable Development Goals" located in the title, abstract, or keywords of the scientific texts. The search including the term "COVID" limited the research to a total of 2 articles, which are automatically included in the scope of the articles, achieved with the research using only the initial two keywords, and these were the selected articles, totalling 40 documents. It was found that not including the term "COVID" among the research keywords, but rather the opportunity to broaden the analysis field, followed the sequence illustrated in Fig. 1.

The period 2015–2020 is justified by the fact that the SDGs were proposed by the UN in 2015. The data used for analysis in this study were collected through

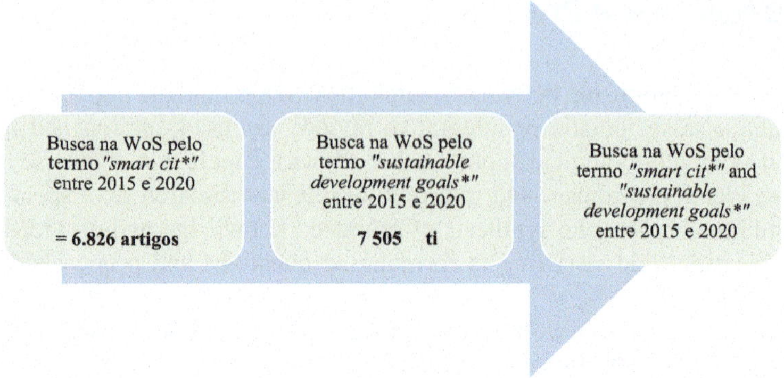

Fig. 1 Filtering design

the Science Citation Expanded (SCI-EXPANDED), Social Sciences Citation Index (SSCI), and the Emerging Sources Citation Index (ESCI) respectively on the Web of Science (WoS). According to Merigó et al. [47], WoS is an internationally recognised digital platform among researchers for offering high-quality standards.

Two bibliometric techniques were combined: performance analysis and scientific mapping. Bibliometric performance analysis uses word frequency analysis, citations, and publication counts by country, university, research group, or authors [72]. Scientific mapping is a spatial representation of how different scientific actors relate to each other [63]. Thus, the objective of this methodology is to show the structural and dynamic aspects of the research.

The content analysis technique [13] enabled understanding the relationship that the analysed themes have in different application contexts and was carried out using the Iramuteq 0.7 Alpha 2 software in qualitative analysis mode. Almico and Faro [6] explain that Iramuteq is a text analysis software that aims to designate the structure and organisation of discourses, informing the relationships between lexical worlds.

Thus, the words used in similar contexts indicate that they are part of the same classification or lexical world, based on a content class derived from vocabulary similarity [7]. In this research, hierarchical descending classification [60] was used, classifying groups with greater predominance of semantic similarity among words, divided by the software with the arrangement of titles, abstracts, and keywords of the 40 articles. With the results generated by this analysis, the researcher has the possibility to reaffirm the direction of the theme in science according to their interpretations.

Fig. 2 Production of papers over the years. *Fonte* Elaborado a partir de dados da *Web of Science*

4 Analysis of Results

4.1 Number of Publications Per Year

Although discussions on the SDGs began in 2015, the year of their creation, their productions related to Smart Cities only had a recorded quantity in this database starting in 2017, as shown in Fig. 2.

Figure 2 demonstrates that the study identified 40 articles on Smart Cities and SDGs in the title, abstract, author keywords, and Keywords Plus® filtered between 2015 and 2020. However, the graph shows that publications dealing with the two terms together only occurred in 2017, with 4 productions. In the subsequent year, it fell to 2 productions, and in 2019, there was an increase in the number of publications within this research scope, being 6 times higher than the previous year. In 2020, the result related to the union of the terms accounted for 22 published articles.

The disposition of these numbers demonstrates the evolution around works related to the theme in academia, suggesting the recognition of the importance of approaches that relate Smart Cities to the fundamentals of the SDGs.

4.2 Most Cited Articles

The productions within the scope of this research cover applications in similar study objects, involving contexts that indicate the direct relationship between the two themes, showing how cities equipped with Smart elements can directly contribute to achieving the SDGs. Moreover, it reveals the debate about indicators as management tools and the importance of their use within the scope of cities and sustainable development (Table 1).

It is worth noting that there is a disparity in the application objects among the most cited articles and some other productions, due to the broad range of applications that the theme offers. Additionally, there are no refinements that segregate the articles by research areas within the filtering criteria, including some articles with technical

Table 1 Most relevant articles from the collected production base—data from Web of Science

Title	Authors	Main theme	Number of citation
Smart and sustainable cities of the future: an extensive interdisciplinary literature review	Bibri and Krogstie [15]	Smart Cities as a requirement for the SDGs	272
Smart Cities: advances in research—an information systems perspective	Ismagilova et al. [32]	Relationship between Smart Cities and the SDGs	103
Don't call me resilient again!: the New Urban Agenda as immunology ... or ... what happens when communities refuse to get vaccinated with Smart Cities' and indicators	Kaika [34]	Smart Cities, SDG 11 and indicators	101
Comparative analysis of standardized indicators for smart sustainable cities: which indicators and standards to use and when?	Huovila et al. [31]	Indicators for Smart Cities with a view to the SDGs	48
Winning the SDG battle in cities: how an integrated information ecosystem can contribute to achieving the sustainable development goals for 2030	Corbett and Mellouli [22]	Information ecosystem to contribute to SDGs	22

content related to the technological area. It is possible to notice a divergence between the study objects but not between the themes.

4.3 Co-authorship Networks

According to Avelar et al. [10], co-authorship networks are employed to investigate existing collaboration patterns among researchers at various levels, such as individual, institutional, national, and international.

Within the scope of this research, the VOSviewer demonstrates that the data collected in WoS from the 40 articles correspond to the work of 126 authors divided into 40 clusters. In this scenario, it is possible to observe that 9 researchers developed their work on the topic individually, however, there are another 31 collaboration networks within the scope of the research (see Fig. 3).

For the purpose of a more meaningful analysis, the research presents nodes that represent articles with a large number of collaborators (six to eight authors), including authors linked to research institutes, government sectors and management and technology companies, distributed in several countries. This analysis demonstrates that the theme goes beyond the fields of science, at a global level, being a growing and dynamic theme, with various contexts of implementation (Fig. 4).

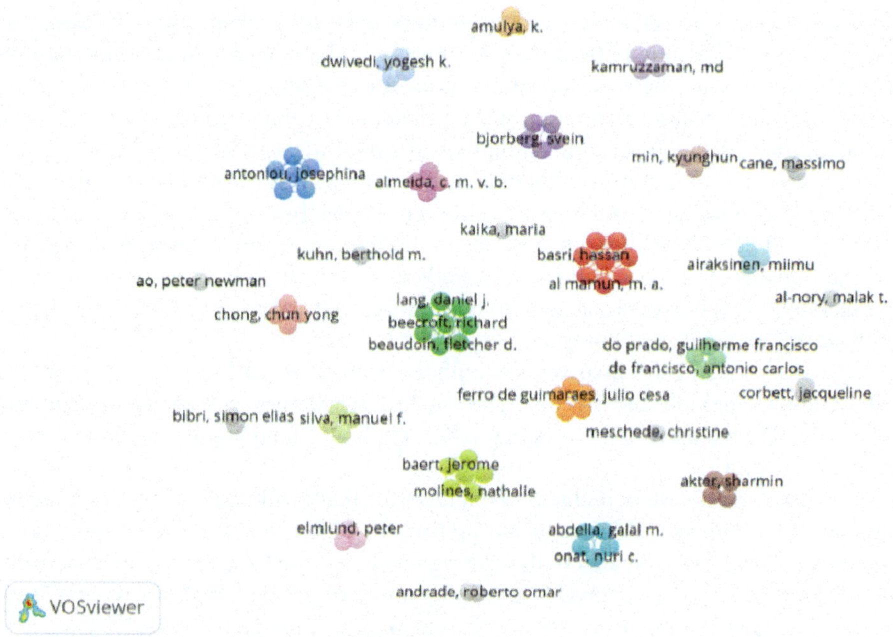

Fig. 3 Co-authorship networks—elaborated using VOSviewer

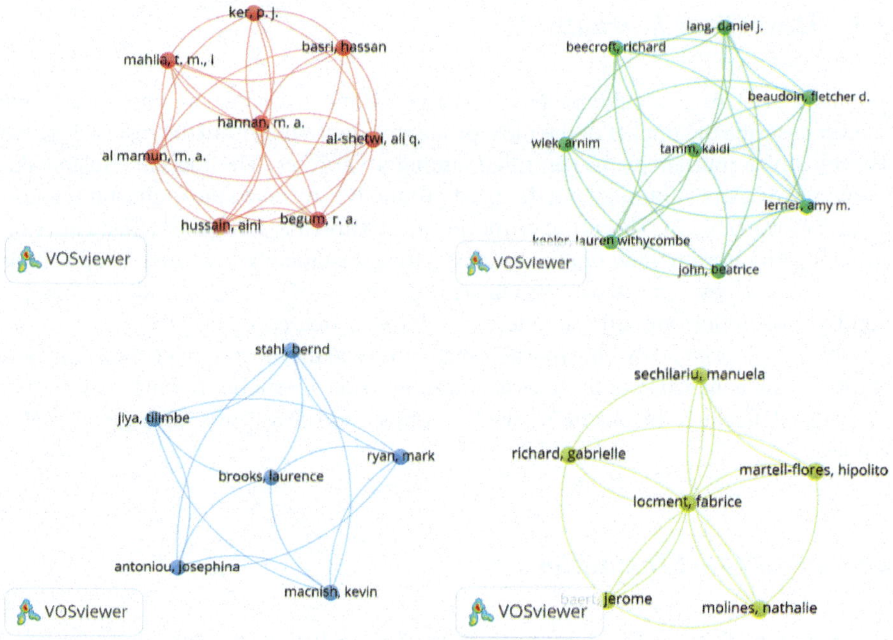

Fig. 4 Authorship networks—elaborated using VOSviewer

The network in red includes eight authors from three institutions in Malaysia and Australia. The authors propose a solid waste collection system in municipalities aimed at improving the service, saving costs, and reducing gas emissions. Furthermore, they demonstrate that the proposed system contributes to the construction of Smart Cities and can be used to strengthen efforts towards achieving the SDGs.

The network in green also includes eight authors from five institutions in the United States, Germany, and Mexico. This network analyses five case studies of partnerships between cities and universities in the three mentioned countries, focusing on the transfer of smart solutions. The authors demonstrate that the city-university partnership can be used to accelerate SDG progress by 2030 and cite Smart Cities as testbeds for smart technologies.

The network in blue involves six authors from five institutions in Cyprus, the Netherlands, England, and Sweden. The article describes how Intelligent Information Systems (IIS) are used to address the SDGs, explicitly focusing on the ethical issue in the process.

The network in yellow includes six authors affiliated with the University of Technology of Compiegne and the consulting firm SYSTRA in France. This study has a technical character, dedicated to describing infrastructure for charging electric vehicles. The production of renewable energy responds to the challenges of Smart Cities, meets the expectations of society, urban planning, and the goals of the SDGs.

4.4 Most Cited Journals

The number of articles published by each journal studied and the number of citations in each describe the impact of the most productive journals within the research scope. Twenty-seven journals were identified among 40 articles. The journal 'Sustainable Cities and Society,' although not the most productive, received the highest number of citations (see Table 2). This is due to the article by Bibri and Krogstie [15], an interdisciplinary literature review that analyses existing sustainable city models and Smart Cities approaches. The debate produces an analysis of strengths and weaknesses, highlighting which strengths are elements that contribute to the SDGs.

However, it is noted that the journals that published the most articles were 'Sustainability,' which has the highest concentration with 7 articles published (15.3%); 'Journal of Cleaner Production,' with 3 articles (11.5%); and 'Energies,' with 3 articles.

4.5 Most Cited Institutions

The number of publications by each institution in the sample and the number of citations each institution has describes the impact of the most productive institutions on the union of SDGs and Smart Cities (see Table 3).

Table 2 Most relevant journals—elaborated using VOSviewer

Journals	Papers	Citation	Total force
Sustainable Cities and Society	2	274	7
International Journal of Information	1	103	3
Environment and Urbanization	1	99	1
Cities	1	48	3
Sustainability	7	40	2
Journal of Cleaner Production	3	30	2
Information Systems Journal	1	22	3
Energies	3	12	0
Urban Science	1	6	0
Applied Sciences-Basel	2	4	0

Table 3 Most relevant institutions—elaborated using VOSviewer

Institutions	Papers	Citations	Total force
NTNU Norwegian University of Science and Technology	2	274	16
FAC Management Law and Social Science	1	103	4
Kerala State Planning Board	1	103	4
Swansea University	1	103	4
University of Amsterdam	1	99	4
University of Manchester	1	99	4
Finnish Association of Civil Engineers RIL	1	48	4
Lapinlahdenkatu 1B	1	48	4
TNO	1	48	4
VTT Technical Research Centre of Finland Ltd	1	22	4

The articles analysed in this study include authors affiliated with 85 institutions. The number of publications by each institution in the sample and the number of citations each institution has describe the impact of the most productive institutions within the scope of this research (see Table 3).

4.6 Country Co-authorship Analysis

Country co-authorship analysis is relevant as it shapes the degree of dissemination of the theme among countries and shows which countries are influential in the studied area. This analysis, within the scope of this research, formed 17 groups. Each cluster is identified with a colour, and the map demonstrates many colours, revealing a diversification in research directions. The larger nodes represent the most influential

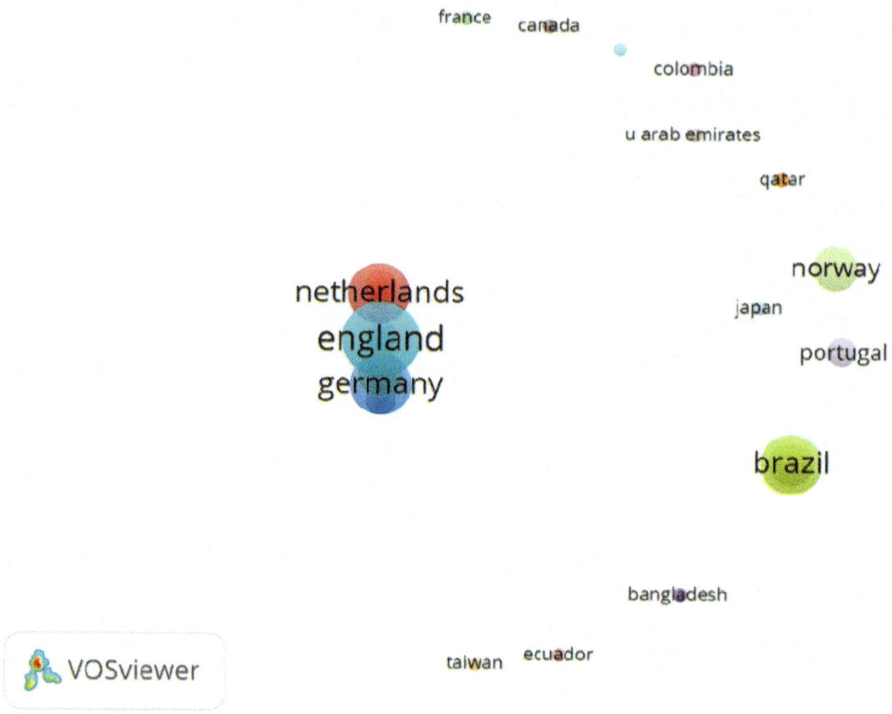

Fig. 5 Co-authorship networks among countries—elaborated by VOSviewer

countries, and the distance between them represents the level of cooperation between countries (see Fig. 5).

According to VOSviewer, the Netherlands, England, Germany, Brazil, and Norway are the most important countries in terms of productions during the period covered by this research, although Norway appears on the map with significant relevance due to being the origin of the most cited article in the research. The analysis shows that although it is a pioneering study scope, the researchers dedicated to this field are spread across 37 countries worldwide and divided into important universities. Furthermore, the productions, even with wide plurality, tend to discuss the aspect that the functioning of Smart Cities, when properly executed, helps achieve the SDGs.

4.7 Key Themes Through Hierarchical Descending Classification

To understand the approaches and main lenses involving the research that relates Smart Cities and SDGs, the Iramuteq software was used, allowing content analyses to identify the most recurring terms, themes, and approaches within this research scope.

For this analysis, the combination of title, abstract, and keywords of the 40 articles was used. An in-depth analysis of the results allowed us to group the various themes into four dominant lenses (classes), which are classified as: (1) Intelligent Processes and Learning; (2) Social Innovation; (3) Sustainability; and (4) Waste Management. These lenses provide a framework that helps to evaluate and understand the literature on the existing interweaving between the study indicators (see Fig. 6).

Very promisingly, the first lens found in Class 1 (34.7%) represents approaches around intelligent processes and learning implemented in cities, transforming the traditional urban environment into an intelligent ecosystem, as suggested by Neirotti et al. [52] and Albino et al. [2]. The COVID-19 pandemic is directly linked to the management of Smart Cities [79]. In this lens, there are two articles that lead the research to the pandemic context due to COVID-19.

The process of learning and adapting to an intelligent ecosystem is the unifying discourse among the analysed articles. From an interdisciplinary review, Bibri and Krogstie [15] address deficiencies in sustainable urban models based on computing

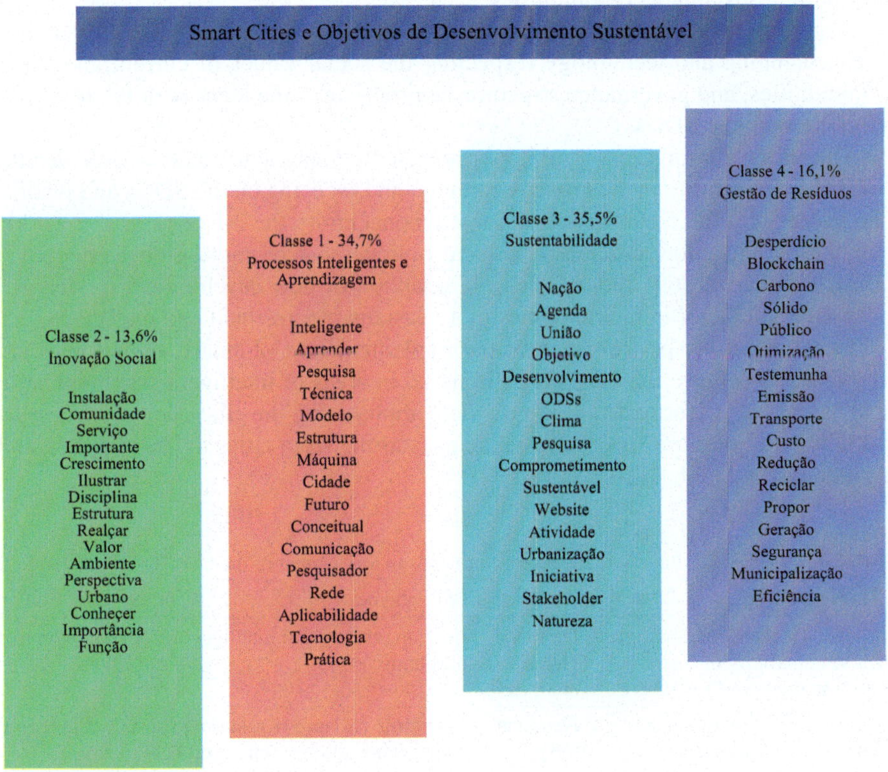

Fig. 6 Descending hierarchical classification

and Big Data technologies. For performance analysis in these environments and visualization of progress in implementing intelligent urban solutions, Huovila et al. [31] propose indicators and models to be used. Digital solutions related to the Internet of Things (IoT) collaborate with the learning process of the circular economy model, being efficient in achieving the SDGs [8]. In this context, within lens 2, there are two productions that address the COVID-19 pandemic (Table 4).

It is possible to infer that the context of the pandemic causes a process of adaptation and change where intelligent processes in city environments remain favorable and constant, managing to overcome the difficulties in achieving the SDGs.

The second lens refers to Class 4 (13.64%) and comprehensively presents themes related to the domains of social innovation, highlighting issues involving infrastructure, regional economic models and governance [14, 45].

Kaika [34] is part of this lens by questioning a change and paradigm in relation to the pursuit of the SDGs by Smart Cities. In relation to economic issues, Mohan et al. [50] present the concept of urban biocycles, which, as in the circular economy, reinvents itself and feeds on the resources generated in the city, being a new production and consumption strategy and, Stanojevic [71] addresses the tourism sector in the Middle West, observing the ability of companies to prosper to achieve the SDGs by building sustainable communities.

Regarding infrastructure, Salaj and Lindkvist [62] and Lindkvist et al. [40] discuss urban planning and technology respecting the social values of current and future communities, and governance structures appropriate to the identity of the locations, aligning with the SDGs.

The third lens of Class 3 (35.5%) represents approaches that involve debates around sustainability in the city environment and focusing on the SDGs and justifies the success by the fact that urban centers concentrate the majority of the world's population and allow efficiency of local policies that promote global effects [5]. Thus, Sotto et al. [69] endorse that the adoption of new development models is a preponderant factor in making cities more sustainable, resilient and intelligent.

Cases about the performance of the implementation of smart technologies and resources to achieve the SDGs are discussed in areas of intelligent communication and aspects of performance measurement, emphasizing the efficiency of actions for sustainability of more specific dimensions, starting from cities to larger scales such as national and global [20, 38, 48].

Table 4 Productions that relate the research themes in the context of the COVID-19 pandemic

Title	Authors	Main theme
Machine learning technologies for sustainability in smart cities in the post-COVID era	De Las Heras et al. [25]	Learning theory and intelligent resources in future problems
COVID, cities and climate: historical precedents and potential transitions to the new economy	Newman [53]	Economic collapse in relation to sustainability and the learning and innovation process to create ecosystems that achieve the SDGs

The fourth lens of Class 4 (16.1%) highlights the main discussion points used to optimize solid waste management in cities and pragmatic actions. Recognizing that the waste disposal and generation system is a recurring and uninterrupted situation in the city environment, the articles in this lens discuss this topic in a more specific way, demonstrating the impact of actions in waste sectors in Smart Cities [52].

Four articles used approaches that demonstrate the importance of waste management actions and policies to achieve greater levels of sustainability, using technological mechanisms to achieve the SDGs. Sechilariu et al. [64] present a smart infrastructure that uses photovoltaic technologies to recharge electric vehicles. Patalas-Maliszewska and Łosyk [56] discuss the implementation and development of low-carbon energy for public transport. Hannan et al. [28] demonstrate a solution for waste collection considering routing through real-time data. Finally, Kang et al. [35] also present a system for collecting garbage using IoT.

On the other hand, two articles discuss the use of blockchain technologies in the process of pollution and urban health, as addressed by Allam and Jones [4]. Pournaras [58] defends the presence of citizens in the decision-making process in relation to crowd detection and the sustainable use of urban transport, using Artificial Intelligence. França et al. [26] propose the application of Blockchain for solid waste management, generating social economic means that contribute to the quality of life in small municipalities and to the achievement of the SDGs.

The analysis of the intertwining of Smart Cities and SDGs and their implications at the time of the pandemic, of the main articles from each of the lenses determined by Iramuteq, is systematized in Table 5. The table allows visualization of the interdependence of the factors worked on in these researches.

Given the complexity and numerous sectors of Smart Cities, as outlined by Neirotti et al. [52], the range of applications involving linkage with the SDGs is varied, but they all cooperate with a common goal: global sustainability. Barber's [11] argument that city governance is crucial to solving global problems and that "mayors govern the world" is highlighted. Intrinsic to sustainability, it is important to recall the social well-being and economic benefits generated by the renewal and optimization of service provision in the urban environment. The global COVID-19 pandemic has demonstrated not only how city sectors can be affected [1] but also the opportunities to build more inclusive, sustainable, and resilient cities by leveraging the potential of technological and information infrastructure in this environment.

5 Final Remarks

The research scope emerged from the observation of a gap in evaluating the linkages related to urban sustainable development. Smart Cities and SDGs have been expanding phenomena in academic studies since 2017 in 37 countries, involving 126 researchers from 85 institutions, as well as governmental sectors and companies. However, in 2020, with the emergence of the COVID-19 pandemic, the need

Table 5 Interweaving of management aspects of Smart Cities to achieve the SDGs and implications at the time of the pandemic

	Authors	Smart Cit*	SDG	COVID 19/ pandemic
Class 1	Bibri and Krogstie [15]	Analysis of Smart Cities approaches	Incorporation of the SDGs in Smart Cities	
	Huovila et al. [31]	Selection of indicators suitable for Smart Cities	Analyzes SDG 11 indicators for Smart Cities	
	Andrade and Yoo [8]	IoT and circular economy	IoT and circular economy: tools for SDGs	
	De Las Heras et al. [25]	Machine learning theory and its applicability	Inclusion on applicability, scopes and dimensions, SDGs to machine learning	Machine learning for Smart Cities and SDGs to face the pandemic
	Newman [53]	Learning and innovation transformations	The economy in urban environments guided by the SDGs	Economic collapse of the pandemic and climate agenda in cities
Class 2	Kaika [34]	Data and environmental monitoring	Meet the SDGs through effective monitoring	
	Mohan et al. [50]	Circular economy and biocycles	Urban economy model aligned with the SDGs	
	Stanojevic [71]	Hospitality and aviation sector in the construction of cities	The debate has practical implications for SDGs 8 and 11	
	Salaj and Lindkvist [62]	Management of urban facilities and participatory governance	Management model that facilitates compliance with the SDGs	
	Lindkvist et al. [40]	Parameters to improve the management of urban facilities	The suggested parameters meet the SDGs	
Class 3	Sotto et al. [69]	Climate issues in urban planning	Local policies to contribute to the SDGs	
	Meschede [48]	Disclosure of information and transparency	Assessment of the SDGs by citizens	

(continued)

Table 5 (continued)

	Authors	Smart Cit*	SDG	COVID 19/ pandemic
	Čepelová and Douša [20]	Implementation of urban sustainability	Analysis of compliance with the SDGs	
	Krzymowski [38]	Sociopolitics and impacts of new technologies	Project implementation analysis	
Class 4	Sechilariu et al. [64]	Energy and transport: charging electric vehicles	The structure serves society and the SDGs	
	Patalas-Maliszewska and Łosyk [56]	Low carbon energy for public transport	Initiative based on the SDGs and public transport users	
	Hannan et al. [28]	Waste collection management by route in real time	This perspective meets 10 of the 17 SDGs	
	Kang et al. [35]	Domestic electronic waste collection system	The solution helps to achieve the SDGs	
	Pournaras [58]	Participatory governance for deliberations using blockchain technology	Participatory governance is aligned with the SDGs	
	França et al. [26]	Blockchain technology, for waste collection and economic generation	The initiative helps to achieve the SDGs	

to adapt this context arose, initiating studies that demonstrate the intensification of emerging elements to provide a new form in the urban world.

Through the conducted review, the study achieves the goal of highlighting the affinity between Smart Cities practices as tools for achieving the SDGs, including the pandemic context. It is found that the relationship between the themes is growing and explores various perspectives on their applications. The inclusion of the pandemic context in this theme takes its first steps in a trend with the potential to advance. The diversity of study implementations in the sectors of Smart Cities linked to the 17 SDGs, including the pandemic context, draws attention to guide future research, such as:

- The evolution of research related to Smart Cities, SDGs, and COVID-19 post-2020.

- Analysis and discussion of key themes of Smart Cities in the COVID-19 context, together or separately.
- Impacts of the COVID-19 pandemic on achieving the 17 SDGs, together or separately.
- Identifying limitations in the existing literature and exploring potential directions for future research.

This study complements area studies, offering an update that should be periodically mapped. The articles analysed here identified, demonstrated, or even measured the direct relationship between the role of Smart Cities in achieving the SDGs and facing the pandemic. The content analysis conducted with the help of Iramuteq and the detailing of relevant articles in each lens demonstrate the usefulness of the resources that characterize a Smart City, proving that, according to theoretical and empirical studies, these resources should be considered central to city management strategies.

The study is limited by having only WoS as a database, covering an initial temporal space concerning COVID-19. Therefore, it is suggested to complement it with research in other databases, including publications from subsequent years. It is important to emphasize that the theme is far from being exhausted and that there is room for more research.

References

1. Abu-Rayash A, Dincer I (2020) Analysis of the electricity demand trends amidst the COVID-19 coronavirus pandemic. Energy Res Soc Sci 68:101682
2. Albino V, Berardi U, Dangelico RM (2015) Smart cities: definitions, dimensions, performance, and initiatives. J Urban Technol 22(1):3–21
3. Aljowder T, Ali M, Kurnia S (2019) Systematic literature review of the smart city maturity model. In: 2019 international conference on innovation and intelligence for informatics, computing, and technologies (3ICT), Sept 2019. IEEE, pp 1–9
4. Allam Z, Jones DS (2020) On the coronavirus (COVID-19) outbreak and the smart city network: universal data sharing standards coupled with artificial intelligence (AI) to benefit urban health monitoring and management. Healthcare 8(1):46
5. Allam Z, Jones DS (2021) Future (post-COVID) digital, smart and sustainable cities in the wake of 6G: digital twins, immersive realities and new urban economies. Land Use Policy 101:105201
6. Almico T, Faro A (2014) Enfrentamento de cuidadores de crianças com câncer em processo de quimioterapia. Psicol Saúde Doenças 15(3):723–737
7. Andrade Júnior EO, Andrade EO (2016) Lexical analysis of the code of medical ethics of the federal council of medicine. Rev Assoc Med Bras 62(2):123–130
8. Andrade RO, Yoo SG (2019) A comprehensive study of the use of LoRa in the development of smart cities. Appl Sci 9(22):4753
9. Araújo CA (2006) Bibliometria: evolução histórica e questões atuais. Em Questão 12(1):11–32
10. Avelar ABA, da Silva-Oliveira KD, da Silva Pereira R (2019) Education for advancing the implementation of the sustainable development goals: a systematic approach. Int J Manag Educ 17(3):100322
11. Barber BR (2013) If mayors ruled the world: dysfunctional nations, rising cities. Yale University Press

12. Barbieri JC, Silva D (2011) Desenvolvimento sustentável e educação ambiental: uma trajetória comum com muitos desafios. Rev Adm Mackenzie São Paulo SP 12(3):51–82
13. Bardin L (2011) Análise de conteúdo. Edições 70, São Paulo
14. Batty M, Axhausen KW, Giannotti F, Pozdnoukhov A, Bazzani A, Wachowicz M et al (2012) Smart cities of the future. Eur Phys J Spec Top 214(1):481–518
15. Bibri SE, Krogstie J (2017) Smart sustainable cities of the future: an extensive interdisciplinary literature review. Sustain Cities Soc 31:183–212
16. Bibri SE (2018) The IoT for smart sustainable cities of the future: an analytical framework for sensor-based big data applications for environmental sustainability. Sustain Cities Soc 38:230–253
17. Buckeridge MS, Philippi Junior A (2020) Ciência e políticas públicas nas cidades: revelações da pandemia da COVID-19. Estud Av 34(99):141–156
18. Caragliu A, Del Bo C, Nijkamp P (2011) Smart cities in Europe. J Urban Technol 18(2):65–82
19. de Castilho Sá M, Miranda L, Canavêz F (2020) Pandemia COVID-19: catástrofe sanitária e psicossocial. Caderno Adm 28(Edição E):27–36
20. Čepelová A, Douša M (2020) Slovakia and the Czech Republic on the path towards sustainable development. Bull Geogr Soc-Econ Ser 47(47):7–25
21. Colding J, Barthel S (2017) An urban ecology critique on the "Smart City" model. J Clean Prod 164:95–101
22. Corbett J, Mellouli S (2017) Winning the SDG battle in cities: how an integrated information ecosystem can contribute to the achievement of the 2030 sustainable development goals. Inf Syst J 27(4):427–461
23. Chechi L, Grisa C (2020) Dos acordos globais às interpretações locais sobre agricultura sustentável: tradução e implementação do Plano e do Programa ABC. Confins Rev Franco-Brésilienne Géogr/Rev Franco-Brasilera Geogr (46)
24. De Jong M, Joss S, Schraven D, Zhan C, Weijnen M (2015) Sustainable–smart–resilient–low carbon–eco–knowledge cities; making sense of a multitude of concepts promoting sustainable urbanization. J Clean Prod 109:25–38
25. De Las Heras A, Luque-Sendra A, Zamora-Polo F (2020) Machine learning technologies for sustainability in smart cities in the post-COVID era. Sustainability 12(22):9320
26. França ASL, Neto JA, Gonçalves RF, Almeida CMVB (2020) Proposing the use of blockchain to improve the solid waste management in small municipalities. J Clean Prod 244:118529
27. Genari D, da Costa LF, Savaris TP, Macke J (2018) Smart Cities e o desenvolvimento sustetável: revisão e perspectivas de pesquisas futuras. Rev Ciênc Adm 20(51):69–85
28. Hannan MA, Begum RA, Al-Shetwi AQ, Ker PJ, Al Mamun MA, Hussain A et al (2020) Waste collection route optimisation model for linking cost saving and emission reduction to achieve sustainable development goals. Sustain Cities Soc 62:102393
29. Harvey D (2001) Spaces of capital: towards a critical geography. Routledge
30. Hollands RG (2008) Will the real smart city please stand up? City 12(3):303–320
31. Huovila A, Bosch P, Airaksinen M (2019) Comparative analysis of standardized indicators for smart sustainable cities: what indicators and standards to use and when? Cities 89:141–153
32. Ismagilova E, Hughes L, Dwivedi YK, Raman KR (2019) Smart cities: advances in research—an information systems perspective. Int J Inf Manage 47:88–100
33. João B, Nascimento D, Souza CLD, Serralvo FA (2019) A systematic review of smart cities and the internet of things as a research topic. Cadernos EBAPE BR 17(4):1115–1130
34. Kaika M (2017) "Don't call me resilient again!": the New Urban Agenda as immunology ... or ... what happens when communities refuse to be vaccinated with "smart cities" and indicators. Environ Urban 29(1):89–102
35. Kang KD, Kang H, Ilankoon IMSK, Chong CY (2020) Electronic waste collection systems using internet of things (IoT): household electronic waste management in Malaysia. J Clean Prod 252:119801
36. Kawakubo S, Murakami S, Ikaga T, Asami Y (2018) Sustainability assessment of cities: SDGs and GHG emissions. Build Res Inf 46(5):528–539

37. Kronemberger DMP (2019) Os desafios da construção dos indicadores ODS globais. Ciênc Cult 71(1):40–45
38. Krzymowski A (2020) Sustainable development goals in Arab Region—United Arab Emirates' case study. Probl Ekorozwoju 15(1)
39. Lazaroiu C, Roscia M, Zaninelli D (2018) Fuzzy logic to improve prosumer experience into a smart city. In: 2018 international conference on smart grid (icSmartGrid), Dec 2018. IEEE, pp 52–57
40. Lindkvist C, Salaj AT, Collins D, Bjørberg S, Haugen TB (2020) Exploring urban facilities management approaches to increase connectivity in smart cities. Facilities
41. Liu H, Li YF (2020) Coronavirus: smart cities could help. Nature 578(7796):515
42. Liu T, Mostafa S, Mohamed S, Nguyen TS (2020) Emerging themes of public-private partnership application in developing smart city projects: a conceptual framework. Built Environ Proj Asset Manage
43. Massuda A, Malik AM, Neto GV, Tasca R, Junior WCF (2021) A resiliência do Sistema Único de Saúde frente à COVID-19. Cadernos EBAPE BR
44. Martins MF, Cândido GA (2008) Índice de Desenvolvimento Sustentável para Municípios (IDSM): metodologia para análise e cálculo do IDSM e classificação dos níveis de sustentabilidade—uma aplicação no Estado da Paraíba. Sebrae, João Pessoa
45. Meijer A, Bolívar MPR (2016) Governing the smart city: a review of the literature on smart urban governance. Int Rev Adm Sci 82(2):392–408
46. Merida C, Lacerda MC (2019) Desenvolvimento sustentável global: "smart cities" e a utilização das novas tecnologias como vetor de cidades sustentáveis. Estud Derecho Iberoamer Vol III:36
47. Merigó JM, Gil-Lafuente AM, Yager RR (2015) An overview of fuzzy research with bibliometric indicators. Appl Soft Comput 27:420–433
48. Meschede C (2019) Information dissemination related to the sustainable development goals on German local governmental websites. Aslib J Inf Manag
49. Min K, Yoon M, Furuya K (2019) A comparison of a smart city's trends in urban planning before and after 2016 through keyword network analysis. Sustainability 11(11):3155
50. Mohan SV, Amulya K, Modestra JA (2020) Urban biocycles—closing metabolic loops for resilient and regenerative ecosystem: a perspective. Bioresour Technol 306:123098
51. NAU Nova Agenda Urbana (2016) Disponível em: https://bit.ly/38uqXWL
52. Neirotti P, De Marco A, Cagliano AC, Mangano G, Scorrano F (2014) Current trends in smart city initiatives: some stylised facts. Cities 38:25–36
53. Newman AO (2020) COVID, cities and climate: historical precedents and potential transitions for the new economy. Urban Sci 4(3):32
54. Nicholas D, Ritchie M (1978) Literature and bibliometrics, vol 29, no 9, pp 423–424. Clive Bingley, London, 183 pp. ISBN 0-208-01541-8
55. ONU Organização das Nações Unidas (2016) Disponível em: https://brasil.un.org/pt-br/sdgs. Acesso em: 01/02/2021
56. Patalas-Maliszewska J, Łosyk H (2020) Analysis of the development and parameters of a public transport system which uses low-carbon energy: the evidence from Poland. Energies 13(21):5779
57. Pautasso E, Osella M, Caroleo B (2019) Addressing the sustainability issue in smart cities: a comprehensive model for evaluating the impacts of electric vehicle diffusion. Systems 7(2):29
58. Pournaras E (2020) Proof of witness presence: blockchain consensus for augmented democracy in smart cities. J Parallel Distrib Comput 145:160–175
59. Rasca S, Waeben J (2019) Sustainable development of small and medium sized cities: use of monitoring frameworks in reaching the SDGs. In: 2019 smart city symposium Prague (SCSP)
60. Reinert M (1990) Alceste une méthodologie d'analyse des données textuelles et une application: Aurelia De Gerard De Nerval. Bull Sociol Methodol/Bull Méthodol Sociol 26(1):24–54
61. Rockström J, Steffen W, Noone K, Persson Å, Chapin FS, Lambin EF et al (2009) A safe operating space for humanity. Nature 461(7263):472–475
62. Salaj AT, Lindkvist CM (2020) Urban facility management. Facilities
63. Small H (1999) Visualizing science by citation mapping. J Am Soc Inf Sci 50(9):799–813

64. Sechilariu M, Molines N, Richard G, Martell-Flores H, Locment F, Baert J (2019) Electromobility framework study: infrastructure and urban planning for EV charging station empowered by PV-based microgrid. IET Electr Syst Transp 9(4):176–185
65. Serey J, Quezada L, Alfaro M, Fuertes G, Ternero R, Gatica G et al (2020) Methodological proposals for the development of services in a smart city: a literature review. Sustainability 12(24):10249
66. Silva ACA, da Silva JF, Santos LRO, Avelino FVSD, dos Santos AMR, Pereira AFM (2016) A segurança do paciente em âmbito hospitalar: revisão integrativa da literatura. Cogitare Enferm 21(5)
67. Silva BN, Khan M, Han K (2018) Towards sustainable smart cities: a review of trends, architectures, components, and open challenges in smart cities. Sustain Cities Soc 38:697–713
68. Sposito M, Guimarães R (2020) Por que a circulação de pessoas tem peso na difusão da pandemia: Difusão da COVID-19 no país segue modelo relacionado a interações espaciais na rede urbana, 26 Mar 2020. UNAN—Unesp Agência de Notícias, Unesp
69. Sotto D, Philippi A, Yigitcanlar T, Kamruzzaman M (2019) Aligning urban policy with climate action in the global south: are Brazilian cities considering climate emergency in local planning practice? Energies 12(18):3418
70. Souza MTD, Silva MDD, Carvalho RD (2010) Revisão integrativa: o que é e como fazer. Einstein (São Paulo) 8(1):102–106
71. Stanojevic S (2020) Embracing sustainability in the drive from ordinary to outstanding: case examples from Emirates, Jumeirah and Rotana. Worldw Hosp Tour Themes
72. Thelwall M (2008) Bibliometrics to webometrics. J Inf Sci 34(4):605–621
73. Tur-Porcar A, Mas-Tur A, Merigó JM, Roig-Tierno N, Watt J (2018) A bibliometric history of the Journal of Psychology between 1936 and 2015. J Psychol 152:199–225
74. UNDESA (2018) World population prospects: 2018 revision of world urbanization prospects. Retrieved from https://population.un.org/wup/Publications/Files/WUP2018-KeyFacts.pdf
75. Van Eck NJ, Waltman L (2010) Software survey: VOSviewer, a computer program for bibliometric mapping. Scientometrics 84(2):523–538
76. Vanti NAP (2002) Da bibliometria à webometria: uma exploração conceitual dos mecanismos utilizados para medir o registro da informação e a difusão do conhecimento. Ciênc Inf 31(2):369–379
77. Vidal JP (2020) Pandemia do COVID-19 e mudanças no Estado: surgirá uma nova administração pública em resposta essas mudanças? Cadernos EBAPE BR 18(4):924–935
78. Washburn D, Sindhu U, Balaouras S, Dines RA, Hayes N, Nelson LE (2009) Helping CIOs understand "smart city" initiatives. Growth 17(2):1–17
79. Won Sonn J, Lee JK (2020) The smart city as time-space cartographer in COVID 19 control: the South Korean strategy and democratic control of surveillance technology. Euras Geogr Econ 1–11

158 documents, by restricting the search to the peer-reviewed articles and book chapters, in English, and in the subjects social sciences, decision sciences, environmental science, engineering, and energy, the sample was reduced to 99 articles. Review of these 99 documents revealed 54 documents that were irrelevant. Reasons for rejection included: no indicators found in the manuscript; the indicators were not related to TODs and or journals had Scientific Journal Ranking below the second quartile. Following the review, two additional articles, cited in the original 99 manuscripts were included in the sample. Resulting in total of 47 documents being included in the investigation. Across the 47 manuscripts, 941 indicators were identified, which were grouped into 115 indicators based on similar terms. However, only indicators with multiple appearances were considered, reducing the final sample to 100 indicators.

2.2 Framework Method

The development of the framework for the classification of sustainability indicators was developed in parallel once the analysis of the results of the literature review was carried out, due to the need to obtain a framework that allows relating the indicators obtained, the pillars of sustainability, the life cycle phase and the actors involved in each phase.

The "*Ecobim Project*" was used as a basis for a relationship between the construction life cycle and the different actors and the "*Sustainability and performance assessment and benchmarking of buildings (SuperBuildings) Research Project*" as a basis for the relationship between indicators and pillars of sustainability through life cycle thinking.

3 Classification According to the Theory

The results of the analysis, shown in Table 1, have been classified following the three most used classification theories: "*3-5-6Ds*", "*Node-Place*" and "*Transit-Oriented-Development*" classification.

The first group follows the characteristics of the "*3Ds*" [12] extended later to "*5Ds*" [23], ("*Density-Diversity-Design + Destination Accessibility-Distance to Transit*"). 5Ds classification was later extended to 6D with the inclusion of "*Demand Management*" [24], but this index was ruled out as no related indicators were found in the conducted analysis.

The second set of indicators were based on "*Node-Place-Performance Theory*" [4]. The third classification utilizes "*Transit*", "*Oriented*" and "*Development*" as indices [56]. The characteristics of the indices from "*Transit*" and "*Development*" are the same as "*Node*" and "*Place*" and the index "*Oriented*" is related to the "*Design*" index from the "*5Ds*" Therefore, it can be considered as an attempt to relate both "*Node-Place Theory*" and "*5Ds*" approaches.

Table 1 Classification of literature according to the theory

Theory and/or classification type	References	Number of references
3 or 5D's [12, 23]	[12, 34, 40, 53, 66, 67, 71, 77, 89, 91]	10
Node-Place [4]	[9, 15, 52, 70, 75, 87, 88, 103, 102]	9
TOD classification [56]	[56, 99, 104]	3
Others	[3, 6, 8, 16, 20, 29, 31, 32, 33, 39, 40–43, 44–47, 61, 62, 64, 68, 81, 85, 98, 100, 105]	25

3.1 Indices and Indicators

Several indicators evaluate different parameters that together form an index [90]. The main characteristics of the TODs, according to the three theories, can be usable as indices to which different indicators are associated. Therefore, "Density", "Diversity", "Design", "Destination Accessibility", "Distance to Transit", "Node", "Place", "Transit", "Oriented" and "Development" are TOD indices. In addition, several indices are assessing one or some of the sustainability pillars of social, environmental, and economic sustainability [18]. The links between different theories and sustainability are used to classify indicators. It is believed that strong relationships exist between the most widely recognised TOD indices, such as "Node-Place", "5Ds", and "TOD" itself. Furthermore, these indices are also linked to the three pillars of sustainability (Fig. 2).

"*Density*" and "*Diversity*" indices are related to "*Place*" and "*Development*" exclusively. The "*Design*" index is mostly related to "*Place*" and "*Development*" although a few indicators are related to transit hub and as such are related to "*Node*" and "*Transit*" indices. The last Ds ("*Distance to Transit*" and "*Destination Accessibility*") are fundamentally bonded to "*Node*" and "*Transit*" indicators.

There are relationships between the economic and social sustainability pillars that combined are considered equitable [94]. "*Distance to Transit*" and "*Destination Accessibility*" are farther away to environmental aspects than other indices. On the other hand, "*Density*" has the strongest connections to environment, with indicators such as green area density or tree density.

Indicators related to "*Design*" depend highly on the values of the design itself and have differing environmental values. Additionally, some "*Diversity*" indicators such

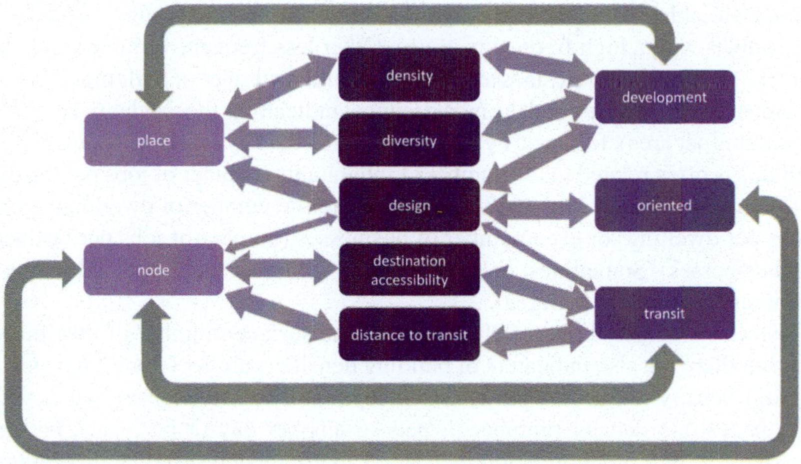

Fig. 2 Relationships between different indices approaches

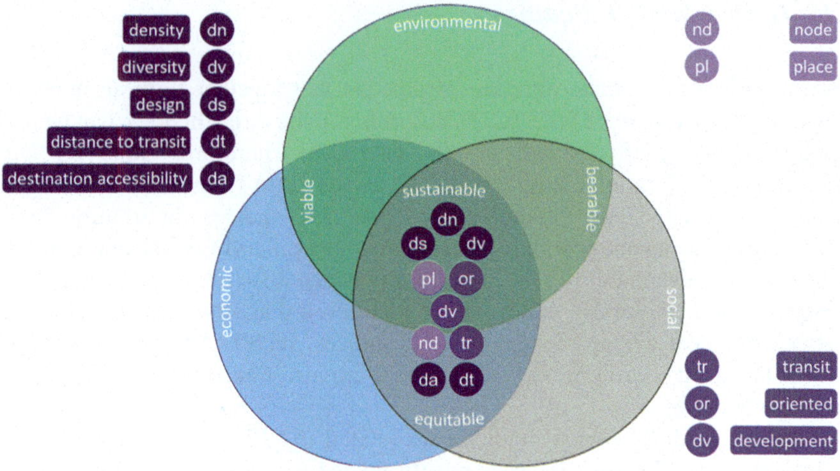

Fig. 3 Relationship between TOD indices and pillars of sustainability

as biodiversity, are also related to environmental issues (Fig. 3). Indicators found in the review have been classified according to the different theories.

3.2 "Density" and "Place" Indicators

"*Density*" is related to two different variables. As shown in Table 2, there are some common variables like inhabitants, area, dwelling number, and number of jobs, which are variables with a high frequency of use. Other less frequent variables include the built area, the number of businesses registered, or the number of buildings. "*Density*" is the most common index and the most frequent indicator is the residential population (inhabitants per area), followed by the occupational density (number of jobs per area), albeit this is often related to the number of inhabitants (number of jobs per inhabitant or 1000 inhabitants). Additionally, it is related to the number of dwellings (number of jobs per dwelling) or even number of businesses (number of jobs per business or 1000 businesses). Sometimes, only commercial businesses are specified (number of commercial businesses per area).

Residential density is generally related to the surface (number of dwellings per area), but there are also indicators of building density (number of buildings per area) and built density (built area per area). Exceptionally, time is also related, such as in the pedestrian density (number of pedestrians per day or per year), but this is more likely to be a monitorization indicator. Environmental sustainability indicators like green area density (green areas per surface) or tree density (number of trees per area) are related by some authors with design. There are also some design related

Table 4 "Design" and "Place" indicators

	Indicator	Unit	"*3/5 Ds*" reference	"*Node-Place*" reference	"*TOD*" reference	Other reference
DS1	Betweenness centrality (BC) or choice value (CV) (Space Syntax)		[89]	[9, 75, 87, 88]	[56]	[98]
DS2	Closeness centrality (CC) (Space Syntax)		[89]	[9, 87, 88]	[56]	
DS3	Eigenvector centrality (EC) (Space Syntax)		[89]	[9]		
DS4	Integration degree (ID) (Space Syntax)			[75]		[32, 98]
DS5	Proportion of four-way intersections	%	[53]			[32]
DS6	Intersection density	%	[12, 34, 53, 66, 67, 71, 91]	[15, 75, 87, 88, 102]	[56, 99]	[41, 45]
DS7	Cul-de-sac density	%	[12]		[56]	[105]
DS8	Motorised vehicle parking places	Number per area, per dwelling, per built are of business use or per habitant	[12, 40, 67, 71]	[9, 102]	[56, 104]	[3, 6, 16, 29, 43, 46, 68, 81]
DS9	Street design	Qualitative	[12, 66]	[9]		[3, 6, 16, 32, 45, 46, 105]
DS10	Street lighting conditions	Lumen (lm)	[12, 77]			[32, 46]

(continued)

Table 4 (continued)

	Indicator	Unit	"*3/5 Ds*" reference	"*Node-Place*" reference	"*TOD*" reference	Other reference
DS11	Priority of vehicles	Priority vehicle passing systems	[12, 77]			[32, 46, 61, 105]
DS12	Paved areas	Paved area per area or in percentage	[71]			[61]
DS13	Sidewalk coverage	% Of streets with sidewalks	[53, 71]	[52]		
DS14	Walkable paths: length and or width	Length per area	[34, 71, 91]	[15, 103, 102]	[56]	[3, 16, 32, 33, 39, 45, 46]
DS15	Street furniture	Units per inhabitant	[71]			[32, 39]
DS16	Equality facilities for pedestrians: handicapped, blinded…	Qualitative				[39, 43]
DS17	Ped shed ratio			[103, 102]		[16, 45, 46, 61, 81]
DS18	Walk score ©			[15, 52, 70, 87]	[56]	[61]
DS19	Bike paths: length and or width	Length per area	[12, 34, 71, 77, 91]	[15, 103]	[56]	[3, 16, 33, 39, 45, 46, 81, 105]
DS20	Number of shared-bicycle stations	Number per area	[89]	[88]		[81]
DS21	Bicycle parking facility	Number per area	[67, 77]	[15]	[56, 99]	[6, 16, 39, 43, 46]

(continued)

Table 4 (continued)

	Indicator	Unit	"*3/5 Ds*" reference	"*Node-Place*" reference	"*TOD*" reference	Other reference
DS22	Quality of bike paths and streetscape	Qualitative	[77]			[46, 105]
DS23	Average block size	Area per area, minimum and maximum length and depth	[12, 53, 71]		[56]	[16]
DS24	Building height	Maximum and minimum height in floors or length	[53, 71]		[56]	[3, 16, 31]
DS25	Infrastructure design: electricity, gas, internet, phone, waste, police…	Qualitative				[43, 64, 68]
DS26	Façade design and transparency	Qualitative	[71]			[16]
DS27	Provision of basic facilities	Qualitative	[67]	[9, 52, 75, 87, 88]	[56, 99]	[39, 43, 46, 61]
DS28	Distance to different facilities: educational, health, commercial…	Maximum in length		[52, 87, 88]		[29, 39]
DS29	Points of interest (POI)	Number per area		[102]	[104]	[32]

(continued)

Table 4 (continued)

	Indicator	Unit	*"3/5 Ds"* reference	*"Node-Place"* reference	*"TOD"* reference	Other reference
DS30	Commercial floor area (CFA) or number of commercial premises	Area per area	[12, 89]	[9]	[56, 99]	[16, 32, 43]
DS31	Mixed use of (first floor) building uses	% Ratio of uses	[71]			[16, 29]
DS32	Housing diversity	% Ratio of types			[56]	[16]
DS33	Distance to parks or open spaces around the site	Maximum in length				[39, 64]
DS34	Open and vacant spaces, and abandoned or vacant buildings	Area per area			[56, 99]	[16, 32, 39, 46, 61, 68]
DS35	Green areas, parks, children's playgrounds and trees	Area per area o per inhabitant	[66, 91]	[52]	[56]	[16, 32, 33, 43, 45, 46, 68]

transportation, even though this last indicator can be monitored through time with crimes per year.

3.6 *Other Indicators*

Other indicators, shown in Table 7, are used to check different variables over a given time. In this study, indicators that are mentioned in at least two investigations have been selected and classified according to the most related sustainability pillar. Therefore, biodiversity loss, air environment quality and air pollution; noise environment quality; water environment quality and re-usement; visual and light environment; energy consumption and CO_2 emissions have been grouped as environmental indicators. Vehicles or distance travelled per capita; level of traffic, pedestrian safety; security in public space and transportation; age and gender group ratio; ethnic ratio and respect given to different social groups; education enrolment and satisfaction with education and participation in neighbourhood organizations and number of organizations have been classified as social indicators and identification as poor; employment rate; employment ratio per sector; tax earnings of district; average income and dispersion inside the TOD; property value and sales; housing affordability as well as housing rent and ratio of rented housing have been categorized in economic indicators. Additionally, not catalogued, weather conditions have been mentioned in some studies.

4 Framework Development

TODS are considered as an example of sustainable urbanism, but little research has been carried out on the relationship between sustainability and urban form of TODs [43, 61]. The vast majority of indicators found are not related to the environmental part of sustainability [41] and only some indicators of environmental sustainability have been found [66].

"*Indicators of Urban Sustainability*" (IUS) are the most widely used tools for evaluating the sustainability of an urban form, but there are many difficulties for its widespread use in any urban form, due, among other factors, to urban complexity, scale, life cycle phase or actor involved in the urban process [18].

"*Transport-Oriented Developments*" and renewals are forms of sustainable urbanism, with a common scale and characteristics which make feasible the challenge of setting a framework of indices and indicators of urban sustainability. However, certain indicators might vary according to the type of TOD as there has been an effort to classify TODs recently according to the urban and regional form and surroundings [41, 80, 101].

Table 7 Monitoring indicators

	Indicator	Main sustainability pillar	"*3/5 Ds*" reference	"*Node-Place*" reference	"*TOD*" reference	Other reference
MO1	Biodiversity loss	Environmental				[16, 61]
MO2	Air environment quality and air pollution	Environmental	[66]			[43, 61, 64]
MO3	Noise environment quality	Environmental	[66]			[43, 64]
MO4	Water environment quality and reusement	Environmental	[66]			[46, 64, 68]
MO5	Visual and light environment	Environmental				[32, 64]
MO6	Energy consumption	Environmental	[66]			[61, 68]
MO7	CO_2 emissions	Environmental				[43, 61, 68]
MO8	Vehicles or distance travelled per capita	Social	[77]		[56]	[16, 46, 100]
MO9	Level of traffic, road network	Social	[66]	[70]		[3, 16, 32, 43, 46, 61]
MO10	Pedestrian safety, accidents per capita	Social				[6, 61]
MO11	Security in public space and transportation	Social				[6, 43, 46, 61, 64]
MO12	Age and gender group ratio	Social	[66, 77]		[56]	
MO13	Ethnic ratio and respect given to different social groups	Social				[61, 64]

(continued)

Table 7 (continued)

	Indicator	Main sustainability pillar	"*3/5 Ds*" reference	"*Node-Place*" reference	"*TOD*" reference	Other reference
MO14	Education enrolment and satisfaction with education	Social	[66]	[88]		[64]
MO15	Participation in neighbourhood organizations and number of organizations	Social				[46, 64]
MO16	Identification as poor	Economic			[56]	[64]
MO17	Employment rate	Economic	[66]	[88]	[56]	[46, 64, 100]
MO18	Employment ratio per sector	Economic	[77]		[56]	
MO19	Tax earnings of district	Economic			[56]	[46]
MO20	Average income and dispersion inside the TOD	Economic	[77]		[56]	[47, 61, 100]
MO21	Property value and sales	Economic	[40]		[56]	[6, 46]
MO22	Housing affordability	Economic			[56]	[46, 61, 64]
MO23	Housing rent, ratio of rented housing	Economic			[56]	[46]
MO24	Weather conditions		[77]			[16, 68]

Most attempts to establish a series of sustainability indicators include frameworks with different indices and indicators, generally grouped according to the sustainability pillar to which they belong [61]. This practice is a simplification of reality, since some indices can influence more than one pillar and different indicators can as well be related one or several indices.

4.1 Framework of Urban Sustainability for Policy Making

To achieve sustainable development and obtain better urban sustainability, "*Life Cycle Thinking*" is necessary [28, 60, 73, 74]. A similar approach is required from all stakeholders involved in the different phases of the construction life cycle, and by extension, the city. Urban policymakers actively participate in two phases, in the brief phase and the operation and maintenance phase [2, 51]. During the brief phase, the sustainability objectives to be obtained are set, whether at a global, regional or local level. The operation and maintenance phase serves to monitor and verify that the sustainable development objectives are being met, and to be able to correct or extend the established sustainability objectives. It is important to set sustainability as a global problem, and it is also appropriate to include some sustainability objectives at a local level, especially in environmental matters. Therefore, when creating any framework for policy making, it is always necessary to provide some flexibility for including objectives at a local or regional level. Our proposal is to provide TOD frameworks as stand-alone units as well as a framework for a series of TODs in a regional level, because each TOD unit might have certain needs and characteristics different from others, but also each TOD must also be related to others, either as a corridor or in a regional level. This framework group is valid for all phases of the life cycle of an urban form, thus allowing the sustainability objectives proposed at the global, regional and local levels to be brought together and to configure the necessary policy regulations to be able to achieve these objectives. It is about making a simple proposal and without endless numbers of indicators of dubious utility [61]. At the same time, it is necessary to include in the framework some monitoring indicators for the operation and maintenance phase, which allows monitoring whether the objectives are being achieved and, in turn, being able to compare some TODs with others (Fig. 4). This will allow a revision of the policy in case objectives are not being accomplished in the future.

4.2 Classification of TODs

To obtain better performance and obtain optimal benefits it is necessary to classify of TODs into different types [41]. Therefore, when analysing the results, and due to the great disparity of results, it is decided to use a practical approach in which the indicators and indices found can be ordered and classified. Approaches

Table 9 (continued)

Pillar	Index	Indicator	Sub-indicator	Life-cycle phase	Urban scale	Actor
		Indicator 3.4.1	Indicator 3.4.1A			
		Indicator 3.4.2	Indicator 3.4.2A			
			Indicator 3.4.2B			
			Indicator 3.4.2C			
			Indicator 3.4.2D			

The "*Catchment Area*" (CA) in a "*Transit-Oriented Development*" can be defined as that residential area around a transport node, which serves as an attraction pole, where the population is able to travel to the node for transit or other daily activities. "*Walking Catchment Area*" (WCA) radius varies between 500 and 1000 m according to previous studies [76]. Beside walking, CA can be based on cycling—electric and regular bicycles, kick bikes—or new transportation methods like shared electrical scooters and bicycles, connected and autonomous vehicles, and mobility-as-a-service. Therefore, it is the area where the majority of the population would be able to go to the public transport station to use it. Hence, there are two limitations, one physical and one psychological, related to the interior mobility.

Walking is established as the only or as the main mobility system inside the TOD in the original definition. Even though Calthorpe mentions certain conditioning factors, such as the climate or the topography, the writing does not present any scientific evidence, not even in the 10 min displacement time; nor in the determined radius of distance of 2000 ft. (609.6 m). Some authors point out that the half mile is, de facto, the standard distance used in the US for planning TODs [30, 69].

Calthorpe's 2000 ft./10 min model equates to a walking speed of 2.3 miles per hour (3.7 km/h). The standardized model of half a mile and 10 min likens a walking speed of 3 miles per hour (4.83 km/h). Pongprasert and Kubota [76] state that 400 m equals 5 min walk and 800 m, 10 min. This statement means a speed of 4.20 km/h. Consequently, in the models there are notable differences regarding the walking speed.

Walking speed is determined by physical conditions, such as age, lower extremity length, strength, weight, gender and spontaneous variability between individuals [65]. Additionally, there are topographical conditions as well as material conditions regarding of the walkable surface [26].

According to the study developed by Gates et al. [27] walking speed is between 3.31 and 4.39 km/h. Therefore, a pedestrian is capable of covering a distance of between 276 and 366 m in 5 min on flat ground, and between 552 and 732 m in 10 min. This result indicates that both values for this indicator have a poor scientific

Table 10 Example of proposed framework for urban sustainability in TOD's though all life cycle phases

Index	Brief phase	Design phase						Main sustainability pillar
		Indicator	Unit		Indicator	Unit		
			Min.	Max.		Min.	Max.	
Index 1	Goal 1.1	Indicator 1.1.1			Indicator 1.1.1A			
					Indicator 1.1.1B			
					Indicator 1.1.1C			
		Indicator 1.1.2			Indicator 1.1.2A			
	Goal 1.2	Indicator 1.2			Indicator 1.2A			
					Indicator 1.2B			
	Goal 1.3	Indicator 1.3.1			Indicator 1.3.1A			
		Indicator 1.3.2			Indicator 1.3.2A			
					Indicator 1.3.2B			
					Indicator 1.3.2C			
Index 2	Goal 2.1	Indicator 2.1			Indicator 2.1A			
	Goal 2.2	Indicator 2.2.1			Indicator 2.2.1A			
					Indicator 2.2.1B			
		Indicator 2.2.2			Indicator 2.2.2A			
		Indicator 2.2.3			Indicator 2.2.3A			
Index 3	Goal 3.1	Indicator 3.1			Indicator 3.1.A			
					Indicator 3.1.B			
	Goal 3.2	Indicator 3.2.1			Indicator 3.2.1A			
		Indicator 3.2.2			Indicator 3.2.2A			

(continued)

Table 10 (continued)

Index	Brief phase	Design phase						Main sustainability pillar
		Indicator	Unit		Indicator	Unit		
			Min.	Max.		Min.	Max.	
	Goal 3.3	Indicator 3.3			Indicator 3.3A			
	Goal 3.4	Indicator 3.4.1			Indicator 3.4.1A			
		Indicator 3.4.2			Indicator 3.4.2A			
					Indicator 3.4.2B			
					Indicator 3.4.2C			
					Indicator 3.4.2D			

base on the distance that most people are willing to walk, and some authors indicate that the distance may differ depending on the person-worker, resident or buyer [30].

Other open discussions are the use of the circle as a delimitation of a scope zone, since this simplification of the space of 10 min can be perfectly delimited today by means of GIS applications, as long as there are some correct calculation parameters. If the Gates parameters are used, real isochrone maps could be perfectly established for distances travelled between 5 and 10 min, with minimum and maximum distances, and more realistic delimitations could be made. In any case, these studies investigate the physical possibility of residents to reach the station in a limited time. The will of the resident should be studied separately, in other issues that have to do with walkability, like the weather, security or others.

Therefore, our proposal is to set "*Indicators of Urban Sustainability*" per "*Walking Catchment Area*" to facilitate policy making and comparisons between different TODs. For example, density thresholds for green areas can be set as square meters per walking CA or entropy can be also de decided per walking CA. Moreover, qualitative indicators are discarded due to the difficulties for comparison in different future case studies. Additionally, quantitative monitoring indicators should be assessed yearly, as some indicators might considerably vary depending on day of the week and/or month or year season.

4.4 Proposed Indices and Indicators

4.4.1 Density Indicators and Residential Capacity (RC)

The main purpose of density is that TODs must have a sufficient population and adequate number of dwellings. Consequently, there should exist a relationship between population density and residential density.

Regarding the appropriate values for each of the indicators, Calthorpe offered in a later publication an indicator for housing density, suggesting 20% of residential use with an average density of "*10–15 dwelling units per acre (25–37 dwelling units per hectare)*" [8] and Cervero between "*20 and 30 dwelling units per acre (50–75 dwelling units per hectare)*" and floor area ratios of 1.0 or above [14].

In any case, a certain number of dwellings does not imply that there will be certain inhabitants, since the dwellings might be empty, used as an investment, as secondary homes or even rented for short stays.

One way of checking the relationship between the number of dwellings and population is through the Residential Capacity (RC) of a territory. A one-bedroom home will usually be inhabited by one or two people, but a three-bedroom home will not necessarily be occupied by one or two people per bedroom as some rooms may be used as study rooms or guest bedrooms, as shown in Fig. 5. In any case, it is possible to obtain a range of minimum and maximum habitability of each dwelling and to be able to carry out studies of the minimum and maximum residential capacity of the WCA or CA of a TOD.

By relating the variables to the pedestrian shed or "*Walking Catchment Area*" (WCA), it is possible to make a better comparison of different case studies. In this way, minimum and maximum thresholds can be established by WCA in inhabitants (habitants per WCA), residential capacity (minimum and maximum RC per WCA),

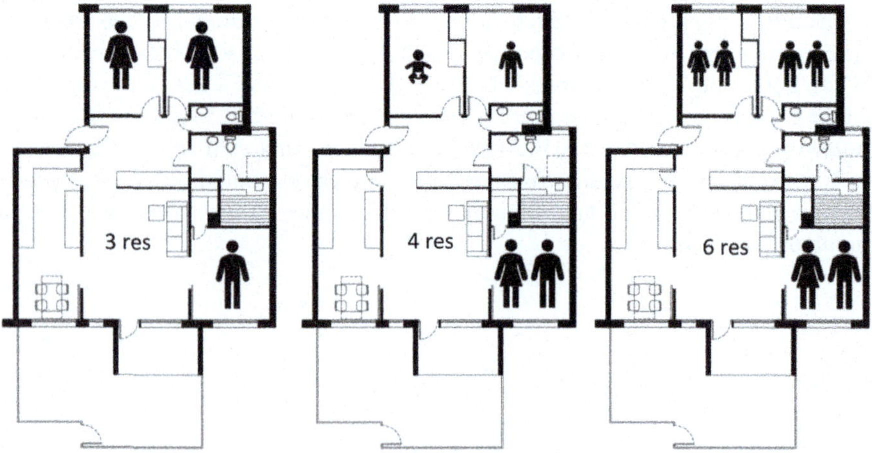

Fig. 5 Example of dwelling with different residential capacity

4.4.4 Distance to Transit and Destination Accessibility Indicators

These indexes are closely related to regional transport planning. Despite policies of regional transport and regional planning must be arranged at the same time as in local level, this scale is out of this review.

5 Conclusion

Despite not being a new concept, scientific interest in TODs has been increasing, especially during the last decade, with many new publications every year [37]. At the same time, the need to achieve an increasingly sustainable development through urban sustainability policies is magnified, due to increasing social awareness of "global social, economic and environmental problems", likewise international instability. Thus, one of the most sought-after goals of urban policy is to achieve sustainable development through urban sustainability practices. In any case, despite there being a multitude of projects in many countries around the world, and being considered a form of urban sustainability, there is a scarcity of TOD sustainability studies that relate urban sustainability, the TOD concept and urban sustainability policies. It is necessary to carry out more in-depth analyses of the urban policy in terms of urban sustainability, as well as a follow-up of the urban areas that allows verifying if the regulations applied help in achieving the established objectives of urban sustainability and development. This chapter aims to address the localized gap by establishing a connection between urban sustainability, urban sustainability policies and the TOD concept through a literary review.

One of the contributions of this study has been the need to incorporate urban life cycle thinking when creating strategic frameworks for urban policy. It is therefore necessary to establish different urban policy frameworks across the various phases of a city's life cycle, rather than focusing solely on the characteristics of the "*Urban Metabolism*". In the brief phase it is necessary to establish different objectives for urban development to be achieved, using different urban sustainability indicators in the design and construction phases that will allow attaining the objectives set in the instruction phase. The construction, commissioning, operation and maintenance phases serve as a monitoring phase for different indicators that permit the verification that the sustainable development objectives are being met, and, additionally, being able to correct or expand the established sustainability objectives if necessary.

The use of the most common characteristics of TODs and relationship with the life cycle is another of the contributions of this research. An approximation through indicators and indices obtained by literary review has been performed, classifying them according to life cycle phase that help create a better vision of small-scale urban policy. A common identification scale in TODs, the WCA, was used, along with indicators with concise units based on that scale, allowing for future comparisons, as each TOD is unique and requires different indicator needs [101], and there are hardly any studies comparing results and indicator units between TODs.

The framework proposed assists in the development of urban sustainability goals and monitoring regulations for TODs. It is focused on the local scale, and, therefore, for standalone TOD units. However, a TOD has independent qualities but at the same time it belongs to a transport corridor, where there are several TODs that also need urban sustainability planning and legislation at the regional level. Therefore, it is necessary to adapt the proposed framework to a regional scale in future research.

Additionally, the units and thresholds of each proposed indicator are pending for further investigations. It is then necessary to carry out case studies that allow a comparison of the units in different TODs, setting a series of minimums and maximums values for each indicator, allowing the achievement of different urban sustainability objectives.

Acknowledgements I extend my gratitude to Steve O'Hern for his valuable supervision and assistance with English language editing.

References

1. Albino V, Berardi U, Dangelico RM (2015) Smart cities: definitions, dimensions, performance, and initiatives. J Urban Technol 22:3–21. https://doi.org/10.1080/10630732.2014.942092
2. Antuña Rozado C, Huovila P, Huovila A, Corredor Ochoa Á (2014) Eco-innovative construction business models for social development. In: Proceedings of XXIV annual RESER conference. Presented at the RESER 2014, Helsinki, pp 72–82
3. Apriansyah D, Barus LS, Zubair A, Simarmata HA, Sumabrata J (2019) Physical analysis conditions around Lebak Bulus MRT station in South Jakarta, Indonesia. Int J GEOMATE 17:63–68. https://doi.org/10.21660/2019.63.8254
4. Bertolini L (1999) Spatial development patterns and public transport: the application of an analytical model in the Netherlands. Plan Pract Res 14:199–210. https://doi.org/10.1080/02697459915724
5. Bianchi M, Menger P, Fernández Fernández I, Maleki P, Rueda F, Fuster Fuengirola E, Etminan G, Margolina A (2022) Methodology for the implementation of a circular economy at the local and regional scale. Publications Office of the European Union
6. Calimente J (2012) Rail integrated communities in Tokyo. J Transp Land Use 5:19–32. https://doi.org/10.5198/jtlu.v5i1.280
7. Calthorpe P (1993) The next American metropolis: ecology, community, and the American dream. Princeton Architectural Press, New York
8. Calthorpe P, Fulton WB (2001) The regional city: planning for the end of sprawl. Island Press, Washington, DC
9. Cao Z, Asakura Y, Tan Z (2020) Coordination between node, place, and ridership: comparing three transit operators in Tokyo. Transp Res Part D Transp Environ 87. https://doi.org/10.1016/j.trd.2020.102518
10. Caragliu A, Del Bo C, Nijkamp P (2011) Smart cities in Europe. J Urban Technol 18:65–82. https://doi.org/10.1080/10630732.2011.601117
11. Carr LJ, Dunsiger SI, Marcus BH (2010) Walk Score™ as a global estimate of neighborhood walkability. Am J Prev Med 39:460–463. https://doi.org/10.1016/j.amepre.2010.07.007
12. Cervero R, Kockelman K (1997) Travel demand and the 3Ds: density, diversity, and design. Transp Res Part D Transp Environ 2:199–219. https://doi.org/10.1016/S1361-9209(97)00009-6

85. Singh YJ, Fard P, Zuidgeest M, Brussel M, van Maarseveen M (2014) Measuring transit oriented development: a spatial multi criteria assessment approach for the City Region Arnhem and Nijmegen. J Transp Geogr 35:130–143. https://doi.org/10.1016/j.jtrangeo.2014.01.014
86. Southworth M (2005) Designing the walkable city. J Urban Plan Dev 246–257
87. Su S, Zhang H, Wang M, Weng M, Kang M (2021) Transit-oriented development (TOD) typologies around metro station areas in urban China: a comparative analysis of five typical megacities for planning implications. J Transp Geogr 90:102939. https://doi.org/10.1016/j.jtrangeo.2020.102939
88. Su S, Zhao C, Zhou H, Li B, Kang M (2022) Unraveling the relative contribution of TOD structural factors to metro ridership: a novel localized modeling approach with implications on spatial planning. J Transp Geogr 100. https://doi.org/10.1016/j.jtrangeo.2022.103308
89. Tamakloe R, Hong J, Tak J (2021) Determinants of transit-oriented development efficiency focusing on an integrated subway, bus and shared-bicycle system: application of Simar-Wilson's two-stage approach. Cities 108. https://doi.org/10.1016/j.cities.2020.102988
90. Tanguay GA, Rajaonson J, Lefebvre JF, Lanoie P (2010) Measuring the sustainability of cities: an analysis of the use of local indicators. Ecol Ind 10:407–418. https://doi.org/10.1016/j.ecolind.2009.07.013
91. Uddin MA, Tamanna T, Adiba S, Bin Kabir S (2023) Revolutionizing TOD planning in a developing country: an objective-weighted framework for measuring nodal TOD index. J Adv Transp 2023:1–25. https://doi.org/10.1155/2023/9911133
92. United Nations (2015a) Resolution adopted by the General Assembly on 25 September 2015—transforming our world: the 2030 agenda for sustainable development
93. United Nations (2015b) Transforming our world—the 2030 agenda for sustainable development (A/RES/70/1)
94. United Nations (1987) Our common future, pp 1–247. https://doi.org/10.4324/9781912281220
95. Wey W-M (2015) Smart growth and transit-oriented development planning in site selection for a new metro transit station in Taipei, Taiwan. Habitat Int 47:158–168. https://doi.org/10.1016/j.habitatint.2015.01.020
96. Wey WM, Chiu YH (2013) Assessing the walkability of pedestrian environment under the transit-oriented development. Habitat Int 38:106–118. https://doi.org/10.1016/j.habitatint.2012.05.004
97. Wey W-M, Hsu J (2014) New urbanism and smart growth: toward achieving a smart National Taipei University District. Habitat Int 42:164–174. https://doi.org/10.1016/j.habitatint.2013.12.001
98. Xu Y, Chen X (2021) Quantitative analysis of spatial vitality and spatial characteristics of urban underground space (UUS) in metro area. Tunnel Undergr Space Technol 111:103875. https://doi.org/10.1016/j.tust.2021.103875
99. Yang L, Song X (2021) TOD typology based on urban renewal: a classification of metro stations for Ningbo City. Urban Rail Transit 7:240–255. https://doi.org/10.1007/s40864-021-00153-8
100. Zamir K, Nasri A, Baghaei B, Mahapatra S, Zhang L (2014) Effects of transit-oriented development on trip generation, distribution, and mode share in Washington, D.C., and Baltimore, Maryland. Transp Res Rec 45–53. https://doi.org/10.3141/2413-05
101. Zemp S, Stauffacher M, Lang DJ, Scholz RW (2011) Classifying railway stations for strategic transport and land use planning: context matters! J Transp Geogr 19:670–679. https://doi.org/10.1016/j.jtrangeo.2010.08.008
102. Zhang Y, Marshall S, Manley E (2019) Network criticality and the node-place-design model: classifying metro station areas in Greater London. J Transp Geogr 79:102485. https://doi.org/10.1016/j.jtrangeo.2019.102485
103. Zhang M, Lee J (Brian) (2023) Make TOD more bicycling-friendly: an extended node-place model incorporating a cycling accessibility index. Buildings 13:1240. https://doi.org/10.3390/buildings13051240

104. Zhou J, Yang Y, Gu P, Yin L, Zhang F, Zhang F, Li D (2019) Can TODness improve (expected) performances of TODs? An exploration facilitated by non-traditional data. Transp Res Part D Transp Environ 74:28–47. https://doi.org/10.1016/j.trd.2019.07.008
105. Zhou J, Yang Y, Webster C (2020) Using big and open data to analyze transit-oriented development: new outcomes and improved attributes. J Am Plann Assoc 86:364–376. https://doi.org/10.1080/01944363.2020.1737182

The manufacturer's authorised representative in the EU is Springer Nature Customer Service Centre GmbH, Europaplatz 3, 69115 Heidelberg, Germany. If you have any concerns regarding our products, please contact ProductSafety@springernature.com

Printed and bound by CPI Group (UK) Ltd, Croydon, CR0 4YY
30/09/2026
02225501-0006